The Last Cavalry Sword

George S. Patton demonstrates 'guard right' with the sword he designed – the US Saber Model 1913. (*Courtesy of the Manuscript Division, Library of Congress, Patton Papers Box 75*)

The Last Cavalry Sword

An Illustrated History of the Twilight Years of Cavalry Swords (UK) General George S. Patton and the US Army's Last Sword (US)

C. Anthony Burke

Frontline Books

First published in Great Britain in 2022 by
Frontline Books
An imprint of
Pen & Sword Books Ltd
Yorkshire – Philadelphia

ISBN 978 1 39908 128 3

A CIP catalogue record for this book is available from the British Library.

Typeset by Mac Style
Printed in the UK by CPI Group (UK) Ltd, Croydon, CR0 4YY.

MIX
Paper | Supporting responsible forestry
FSC® C013604

Pen & Sword Books Limited incorporates the imprints of Atlas, Archaeology, Aviation, Discovery, Family History, Fiction, History, Maritime, Military, Military Classics, Politics, Select, Transport, True Crime, Air World, Frontline Publishing, Leo Cooper, Remember When, Seaforth Publishing, The Praetorian Press, Wharncliffe Local History, Wharncliffe Transport, Wharncliffe True Crime and White Owl.

For a complete list of Pen & Sword titles please contact

PEN & SWORD BOOKS LIMITED
47 Church Street, Barnsley, South Yorkshire, S70 2AS, England
E-mail: enquiries@pen-and-sword.co.uk
Website: www.pen-and-sword.co.uk

Or

PEN AND SWORD BOOKS
1950 Lawrence Rd, Havertown, PA 19083, USA
E-mail: Uspen-and-sword@casematepublishers.com
Website: www.penandswordbooks.com

Contents

Acknowledgements

Without the kind assistance of the following individuals this book would not have been possible, and the author would especially like to thank Mr David Miller, Curator, Military History, at the Smithsonian Institution National Museum of American History, who during several photographic sessions made the institution's extensive sword collection available, answered numerous questions, and provided every conceivable assistance. Four other curators provided valuable assistance: Alex MacKenzie of the Springfield Armory Museum, who provided images of unique prototypes of US sabers only extant in that collection; Kaleb Bemis at the Museum Resource Center of the Rock Island Arsenal, who provided the 1913 Springfield Armory technical drawings used to manufacture the saber; Neil Abelsma at the National Museum of the US Army, who provided images of the first Model 1913; and Nathan Jones at the General George S. Patton Museum, who provided images of the prototype Model 1913 that was the personal property of George S. Patton. The author is also grateful to the staff at the National Archives and the Manuscript Division of the Library of Congress, especially Lewis Wyman, who provided research assistance.

In addition, the author would like to thank Tom Graham and Roy Fogle, who made their private collections available, as well as their time and knowledge.

Collectors should be aware of the following books, which in turn made this book possible and which the author highly recommends:

- Crouch, Howard R., *Historic American Swords*, 1999, SCS Publications.
- Farrington, Dusan P., *Sword & Sabers of the United States Army, 1867–1918, The Regulation Models*, Woonsocket, Rhode Island, Mowbray Publishing, 2015.
- Peterson, Harold, *The American Sword 1775–1945*, Philadelphia, Ray Biling Arms Books, 1991.

For further sources, readers should consult the Bibliography.

Introduction

Arma virumque cano – 'I sing of arms and the man' is the first line of *The Aeneid*, Virgil's epic poem about the hero Aeneas and his deeds. And one could begin similarly: I write about an American soldier and the sword he made, the last sword made to serve the army of his country. The sword, the US Model 1913 Cavalry Saber, the soldier, George S. Patton, who after directing the modernization of the sword, a weapon used since antiquity, would master a thoroughly modern weapon, the tank, first as an instructor and leader in the First World War and then as the general commanding the Third Army to victory in the Second World War.

The ancient Romans may have formed the first armies on a scale similar to those of the modern world and the first large-scale armies that were uniformly equipped. Therefore, the first model infantry sword was the Roman short sword, the gladius, and the first cavalry sword, the spatha – the longer Roman cavalry sword. These examples from ancient Rome may have been the first, but the US Cavalry Saber and Scabbard Model 1913 – the Patton saber – was the last.

Some famous weapons made their designers equally famous and the weapons are known not by a model and year, but by the names of their creators: in the nineteenth century 'Colts' and 'Winchesters' for men who revolutionized the revolver and repeating rifle and whose names became synonymous with their designs. Similarly, in the twentieth century there were men who designed weapons that would make their names well known: 'Luger' for the man who improved Borchardt's design and created the Pistolen 08; 'Kalashnikov' named after the designer of one of the most important assault rifles of the modern world, the AK-47; and 'Garand' whose M-1 rifle was used by the United States and allies during the Second World War and Korean War.

The names of these men are known to history solely because of the weapons they designed. Not so with Patton. It was not the design of the final combat sword of the US Army that made Patton famous; his fame was determined by his battlefield success as a commander of armies during the Second World War. However, that success was toward the end of a lifetime devoted to the profession of arms, and it should be noted that designing the last cavalry sword of the US Army was an important part of his development as an army officer.

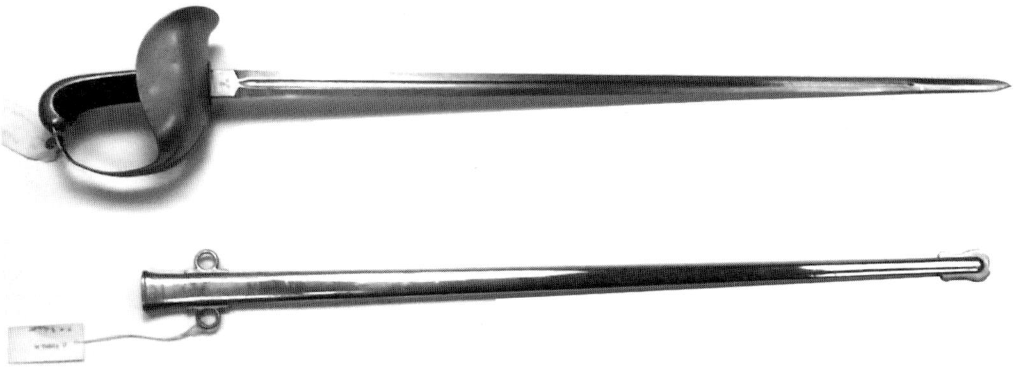

Figure 1. This Cavalry Saber and Scabbard Model 1913 (Officers' Model) was the personal sword of General Paul Robinett, who like the sword's creator, George S. Patton, was a US Army cavalry officer, Olympic athlete, and student at the French cavalry school at Saumur. Robinett carried this sword as a member of the US Cavalry Team during the 1920s, made a private purchase of the sword and donated it to the Smithsonian Institution in 1969. Robinett served as a general officer in the North African campaign where he was wounded in action. Because of the severity of his wounds, he served out the rest of the war at Fort Knox, Tennessee. (*Courtesy of the Smithsonian Institution, National Museum of American History*)

Finally, the sword Patton designed was the last sword developed by a major power **intended for use as a weapon** by soldiers of its army. The final German cavalry saber was the Prussian Model 1889; for France it was the Model 1896; Italy and Sweden would also develop their final cavalry sabers at the end of the nineteenth century. Spain would issue its final cavalry saber in 1907, Great Britain in 1908, with a lighter nickel-plated officers' version in 1912.

How did it happen that in the year before the First World War – a war marked by the first use of many modern weapons, most decisively, the machine gun, the tank, and fighter and bomber aircraft – the US Army Ordnance Department developed a new cavalry sword: the Cavalry Saber Model 1913. Although the idea of mounted cavalry fighting with swords may seem anachronistic to our conception of the state of things in 1913, it was common military practice for modern armies to have mounted cavalry armed with swords.[1]

In the 1932 catalog of the swords in the United States National Museum of the Smithsonian Institution, curator Theodore Belote described the Model 1913 as an 'entirely new type of saber.'[2]

Harold Peterson, the foremost American authority on US swords, described the Patton cavalry sword as perhaps the best sword ever issued to cavalrymen serving in the US Army. In his book *The American Sword 1775–1945*, he wrote that the beginning of the twentieth century would see the last cavalry charge of the US Army with sabers; the Americans who made that charge did so with sabers left over from the Civil War. However, as Peterson explained, the desire to update the Civil-War era saber had begun: '… agitation was already powerful for a new and more efficient model and in 1913 such a sword was adopted … perhaps the best that was ever carried by an American cavalryman.'[3] It should be noted that US cavalry troopers in the First World War were equipped with the Cavalry Saber Model 1913 (see Figures 68 and 69 below) and they may have been used in combat, even though there were no documented cases.

A complete history of swords would comprise a number of volumes and is not the purpose of the author. Although an overview of the history of swords is necessary, the focus will be on cavalry sabers, because unlike the infantry sword, the cavalry saber functioned as an important weapon from the beginning of English colonization in America up to the time that Patton designed the US Army's last cavalry saber. Rather than provide a complete history of swords, this book will describe the evolution of light and heavy cavalry swords in Europe up to the early twentieth century, which influenced sword design and production in the United States.

The book will then provide a brief discussion of swords in colonial North America, the early republic and, in more detail, of US cavalry swords from the Civil War to the period just before the First World War, and conclude with the creation of the Cavalry Saber Model 1913 designed by George S. Patton, who had unique qualifications to be its creator. This book describes the excellence of the Patton saber, which reflects Patton's own single-minded pursuit of excellence in the profession of arms. However, first, a brief overview on the subject of the sword is in order.

Some Definitions: The Sword and the Saber

Just as all squares are rectangles, but not all rectangles are squares, so all sabers are swords, but not all swords are sabers.* In his *The Book of the Sword*, the English explorer Sir Richard Burton defined a sword as follows:

> Physically considered, the sword is a metal blade intended for cutting, thrusting, or cut-and-thrust. It is usually, but not always, composed of two parts. The first and principal is the blade proper … Its cutting surface is called the edge

---

* Note to the reader – throughout this book the American English word 'saber' will be used except when quoting a source that uses the alternate form 'sabre'. In addition, some sources are quoted using the word 'gripe' as an alternative for the word '"grip'. The word 'sword' is Germanic originating from the Old English 'sweord'. The Latin word for sword is 'gladius', which only appears in modern usage in the word 'gladiator'. However, importantly, the Latin word for the Roman cavalry sword, 'spatha', is the root of many modern Romance language words for sword: in Spanish, 'espada' and in French, 'épée.'

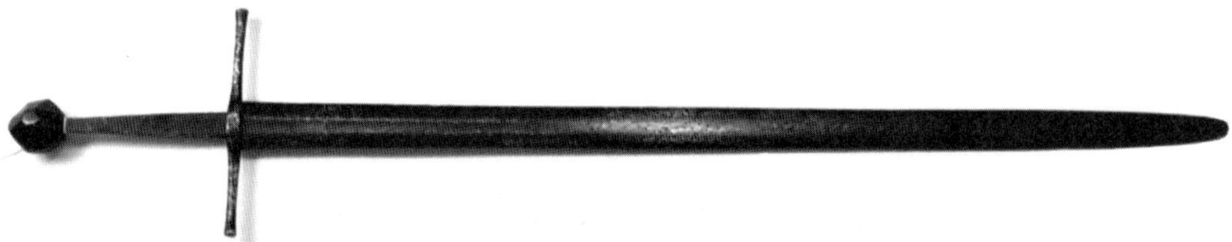

Figure 2. Late medieval sword, English, fourteenth century, discovered in Cyprus, probably used during the Crusades. (*Courtesy of the Smithsonian Institution, National Museum of American History*)

> … and its thrusting end is the point … The second part, which adapts the weapon for the readier use, is the hilt.[4]

A saber, however, is a specialized sword, usually with a curved blade for cutting and designed for use with one hand. The ordnance departments of some armies, however, have described any sword with a hilt commonly seen on a saber as a saber, regardless of the type of blade.

Figure 3. Early eastern or central European saber. (*Courtesy of the Smithsonian Institution, National Museum of American History*)

Chapter 1

Swords, from the Roman Cavalry to the Medieval Knights

Throughout most of history, the sword was an ancillary weapon. In heroic Greece, Homer's heroes fight each other with shield and infantry spear. Caesar's legions led with the pilum and followed up with the gladius, the Roman short sword, while they were protected by the scutum – the large rectangular shield.

By the second century AD, the Roman infantry followed the cavalry in adopting a longer sword, the spatha, which could be almost 8in longer than the gladius.[1] This longer sword is one of the ancestors of the swords later carried by European knights.

During the Dark Ages that followed the disintegration of Roman Imperial power in the West, European smiths pattern-welded sword blades to overcome the difficulties of producing homogeneous steel blades of high enough quality to remain sharp, but not crack, while using both inferior and good-quality iron.[2]

> A pattern-welded blade of the highest quality could take a month to make … Such blades were made of many separate parts; a centre bar of complex structure forged into one, and two edges made from a long billet of relatively homogenous steel bent back on itself into a tight 'V', which was then hammer-welded into place along each side of the center bar to give the blade a hard cutting edge and point. To make the centre, several thin rods of malleable wrought iron were first case-hardened in a charcoal fire. Through this process the iron on the surface was carburized (that is, it absorbed carbon) to form a skin of hard steel. The rods were then heated red-hot and tightly twisted together, and finally were hammer-welded at white heat to forge all the constituent parts together. Another important ancestor was the sword of the Vikings, who developed a technology that made pattern-welding bits of iron and steel obsolete. The Vikings fabricated long swords of homogeneous steel that equaled or exceeded the quality of the pattern welded blades.[3]

Knights at the end of the twelfth century could and would use the point of their swords, as the painting *The Victory of Humility over Pride* detail from the Trier Jungfrauenspiegel, in the Kestner Museum, Hanover shows.

The stirrup enabled Frankish and Norman horseman to focus the weight of man and horse in the sharp point of the lance; so it's not surprising that in Shakespeare's *The Life of Henry V*, Pistol asks the incognito King Henry V, 'Trail'st thou the puissant pike?' or, as we would say today, 'Are you armed with a lance?' not 'Do you carry a sword?'* For as exalted a weapon as the sword was, the lance was the primary weapon of the knight, just as the recurved bow and arrows or javelins were the primary weapons of Saladin's warriors, whose sabers, like the knights' swords, were ancillary weapons, albeit important ones.

Perhaps the most important development of the Dark Ages was that the knight, supported by his stirrups, could deliver a more powerful cut with his sword than a man without stirrups.

By the First Crusade, a Norman knight wore much the same armor as the prior generation wore at the Battle of Hastings. The knight who took part in the conquest of Jerusalem was covered from his shoulders to his calves by a mail shirt, iron or steel rings on a leather foundation, a mail coif protected his neck and throat, and he wore a conical steel helmet with a vertical flange to screen his nose. He also carried a kite-shaped shield that protected his left side, his left arm secured the shield while his left hand held the reins leaving his right hand free to wield lance and sword, for from his left side hung a scabbard containing a double-edged sword.[4] A hundred years later, a knight who was among those defeated at Hattin and would see Jerusalem fall to Saladin would have been recognizable to his peers by the armorial devices on his shield and surcoat. Also, the armor he wore was more skillfully made and comprised mail leggings as well as a tighter fitting mail upper shirt. Compared with his grandsires, it was like the difference between a man in a tailor-made suit and a man wearing a scarf and greatcoat.[5]

By 1350, knights who could afford it were, for the most part, armored in a suit of brightly polished steel plates, hard enough to deflect the bolts fired by infantry armed with crossbows, and shiny enough to create the image of 'a knight in shining armor.'[6] These suits were tailored to fit the wearer, who could perform surprising feats of arms in them, because, unlike mail, the armor did not pull down on the wearer, like a heavy oversized, waterlogged sweater but was supported by a series of straps and braces, which evenly distributed the weight.

Because of the knight's armor and shield, his sword was primarily an offensive weapon in the main used on horseback with one hand; the sword usually had a simple guard to protect the hand and a total length of 35 to 40in. The blades of these swords were flat, as they were used generally for cutting, but beginning in the thirteenth century and continuing in the fourteenth century, swords were made with sharply pointed narrower blades, with better rigidity for thrusting, probably

* William Shakespeare, *The Life of Henry V*, Act 4, sc. i.

because of the greater use of plate armor. Although often shown in movies and popular culture to be ponderous, the knight's sword weighed about the same as a cavalry trooper's saber.

By the fourteenth century, a large type with a total length of 45 to 55in was also used.[7] These hand-and-half swords would be employed by the rank and file cuirassiers of the fifteenth and sixteenth centuries.

During the Middle Ages, there was a considerable class division on the field of battle, the upper classes, knights, and peers fought on horseback with lance and sword; infantry levies were mostly composed of peasants armed with warlike versions of their daily implements, flails, and polearms made from scythes and pruning hooks. But after the Middle Ages, the knight with his heavy armor and lance would fall victim to the musket in the same way that the castle would fall victim to the cannon. However, during all these developments in warfare, the sword would increase in importance, particularly among cavalry, and the descendants of the knights, the cuirassiers, would be an important component of armies from the age of Gustavus Adolphus to the Napoleonic era.

Chapter 2

Overview of Cavalry Swords and Cavalry Sabers and their Development in Europe and Great Britain

As the founders of the United States were of English descent and their military advisors were mostly from France, various German states, or other nations on the continent, the Continental Army was organized like the armies of Great Britain and Western Europe with three major branches: infantry, artillery, and cavalry, each using weapons either directly imported from Great Britain and Europe or closely based on those weapons. Therefore, a brief overview of the role of these branches in European armies and their weapons contemporaneous with the colonization of North America and the founding of the United States is in order.

New methods and materials of warfare appeared more frequently in Europe than in Asia because there was no exclusive control of the technology of warfare as there was in Mogul India or Tokugawa Japan, where the overlords exercised complete dominance over the science of firearms.[1] Further the rise of the independent city-states and the mercantile economy of late medieval and early modern Europe brought an end to the dominance on the battlefield of the knight and his ill-trained feudal levy. Whether it was companies of Italian crossbowmen protected by compatriot pikeman or Swiss pikemen or Spanish halberdiers, the rulers no longer relied on feudal levies but on armies of professional mercenaries who expected payment. Certainly, many knights found service among the companies of heavy cavalry or even as leaders of these companies.

By the Renaissance, professional armies divided their labor on the battlefield through specialized branches: infantry, artillery, and cavalry.

After the Middle Ages, as the feudal system lost political power in Europe to cities and princes, the large masses of inefficient foot soldiers were replaced by hired mercenaries, with money and recruiting being of great importance from the Thirty Years War to the wars of Louis XIV.[2]

Infantry

Originally comprising just pikemen or halberdiers, infantry was later divided into pikemen protecting crossbowmen or musketeers; but the invention of the plug bayonet, a dagger with a handle that fitted into the muzzle of the musket, began to make the pike obsolete. Further development of the musket from a clumsy, heavy matchlock, which was so heavy that it was supported by a rest, to a slender well-balanced flintlock, and the perfection of the bayonet so that it no longer plugged the muzzle but was held onto the barrel with a ring and retaining lug, enabled the infantryman to fix bayonet and still fire his musket, greatly enhancing the role of the infantryman on the battlefield while standardizing his appearance.

Artillery

During this time, the purpose of artillery advanced from the highly specialized siege-warfare task of demolishing castles and fortresses to a significant role on the battlefield of breaking up infantry squares and the lines of charging cavalry. It is important to note that there was little difference between the cannons used in Europe and Asia in the mid-fifteenth century. But the competitive nature of European power struggles provided a drive for constant improvements in gunpowder grains and cannon castings. The French who invaded Italy in 1494 with powerful bronze cannons, which began an arms race between the cannon makers and the fortification designers, found that raised earthwork fortifications – compacted mounds of earth with steep ditches at the front and a sophisticated series of protected bastions – provided protection against armies with artillery trains.[3]

A good sense of the effect of artillery in Europe between the late Middle Ages and the Napoleonic era is provided by Carl von Clausewitz, who served as an officer in the Prussian Army during the Napoleonic Wars and whose book, *On War*, is one of the most insightful works on the nature of war ever written:

> In the Middle Ages firearms were a new invention, so crude that their physical effect was much less important than today; but their psychological impact was considerably greater. One has to have seen the steadfastness of one of the forces trained and led by Bonaparte in the course of his conquests – seen them under fierce and unrelenting fire – to get some sense of what can be accomplished by troops steeled by long experience of danger, in whom a proud record of victories has instilled the noble principle of placing the highest demands on themselves. As an idea alone it is unbelievable. On the other hand, there are European armies that still have troops such as Tartars, Cossacks, and Croats whose ranks can easily be scattered by a few rounds of artillery.[4]

The types of fortifications designed in early modern Europe using thick mounds of earth to defend against the advances in artillery were brought to the Americas by European explorers and colonists; they can be seen in the few remaining earthwork fortifications left over from the American Civil War. As artillery became more mobile, particularly horse artillery, many mounted artillerymen were lightly armed as their cannons were their primary duty and they were armed with swords similar to those carried by light cavalry.[5]

Light and Heavy Cavalry in the Renaissance – the Saber Makes its Entrance

Medieval cavalry in Europe was generally composed solely of fully armored knights. It is incorrect to think that during the Middle Ages there were more mounted soldiers than infantry; it's not that there were more cavalrymen, in fact, the infantry levies were much more numerous, however, cavalry was more effective, but that would begin to change as the Middle Ages drew to a close.[6] 'The custom of using large masses of relatively inefficient foot soldiers came to an end only when the feudal system of military service was replaced by that of hired mercenaries, and the conduct of war became dependent on money and recruiting.'[7] As infantry and artillery became more effective, cavalry had to adapt, and threats to Europe from the East provided the impetus to develop new forms of cavalry.

By the late fifteenth century the Ottoman Turks, who had conquered the Eastern Roman Empire and put Constantinople to the sword, were beginning to make war on the periphery of the Austrian Empire and their light cavalry was effective against the European defenders. While it is true that European Crusaders employed lightly armed cavalry, Turcophiles, against the Saracens, in Europe, for centuries, cavalry was armored and elite.

The Holy Roman Emperor, Maximilian I, and his successor, Ferdinand I, devoted considerable time and resources toward defense, twice repelling the Ottomans from their attempts to seize Vienna. After the subsequent devastation in the border country of Styria, the Habsburgs built fortifications to withstand the artillery of the period.

The Styrian cavalry was divided into two types of light cavalry and one type of heavy cavalry:

- The harquebusier riders armed with two to three wheel-lock pistols and armored in a burgonet, a bulletproof cuirass with mail sleeves, had the mobility and firepower to best combat the Ottoman horsemen.
- The hussars, armed with lance, estoc, and saber, were less heavily armed than the harquebusier riders, they wore a laminated cuirass and mail shirt.
- Heavy cavalry, cuirassiers, wore three-quarter armor, including a close helmet, cuirass, plate arm and leg protection to the knee, and heavy riding boots. They were armed with straight-bladed hand-and-half swords and pistols.[8]

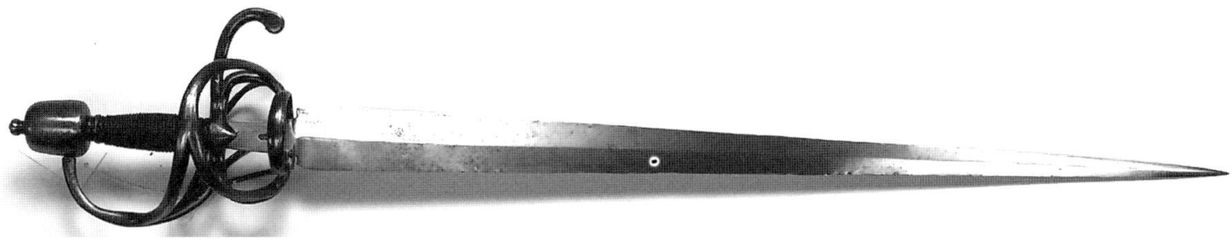

Figure 4. Swept-hilt rapier with a robust blade, meant for the battlefield, not the streets of a Renaissance city. The blade was probably made in Toledo, Spain, whereas the hilt may have been made in Germany. (*Courtesy of the Smithsonian Institution, National Museum of American History*)

The heavy cavalry trooper of the Renaissance, armored as described above and armed with wheel-lock pistols, sometimes a heavy lance, and a long robust double-edged sword (some of the officers carried robust rapiers – see Figure 4 above), was the descendant of the fully armored knight and the predecessor of the Napoleonic cuirassier. Regarding the use of robust rapiers in war, it should be remembered that the Spanish cavalry continued to use rapiers until the end of the eighteenth century.[9] From the Renaissance to the Napoleonic era, heavy cavalry troopers charged the enemy to provide a tremendous shock in the height of battle to overwhelm infantry squares.

Contacts similar to that described above between the Ottoman and the Austrians occurred along much of Europe's eastern marches, and Poles, Croats, and Hungarians, similarly, through contact with the Near East, began to emulate the light cavalry of the Turks in equipment and tactics. Turkish and other eastern light cavalry were primarily horse archers, who, for close combat, were armed with a saber and small shield, as their sharply curved sabers were not ideal for parrying, but were excellent for the close-in fighting of light cavalry engagements. In these sudden melees where horsemen were closely pressed, giving point would be difficult and consequently light cavalrymen of both East and West used sabers.

Armed with sabers, the Hungarian and Polish cavalry succeeded in the wars against the Turks, most importantly by foiling the attempted siege of Vienna in 1683; as a result, beginning with the French and Bavarians, other nations began to adopt the saber as the standard weapon of light cavalry.[10]

> [The early saber] had a broad, single-edged blade, usually curved, a cap-shaped pommel, a short lobe-shaped rear quilon which, after bending down at right angles, continued to the pommel as a knuckle-guard. This shape of hilt, called a stirrup-hilt in England, remained in use on the light-cavalry sabre until the second quarter of the nineteenth century. Light swords with stirrup-hilts were also used by certain infantry and artillery units, including the Light Infantry and Grenadier companies of English line regiments, during the second half of the eighteenth and early nineteenth centuries.[11]

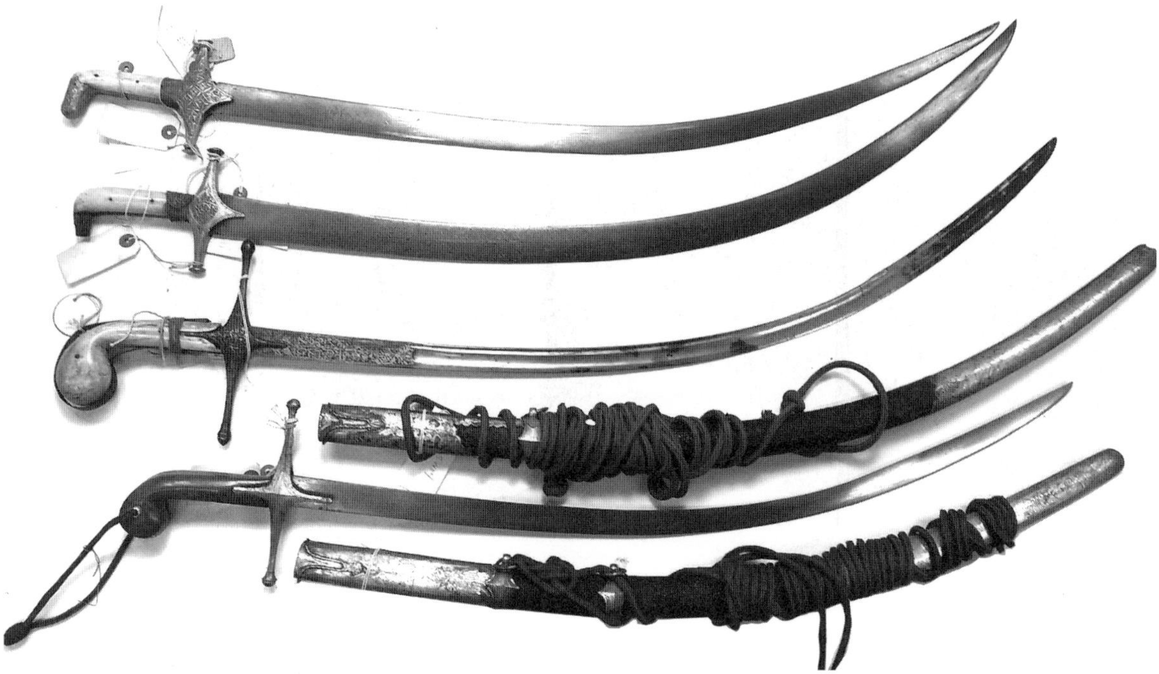

Figure 5. Four Near Eastern sabers in the collection of the Smithsonian Institution. (*Courtesy of the Smithsonian, National Museum of American History*)

Perhaps the best authority on swords and swordsmanship to describe the advantages of a curved sword was the explorer Sir Richard Burton, in his *The Book of the Sword*. Although he gave a compelling argument for the overall superiority of straight swords, as we shall see later, he explained why curved swords may be better at cutting:

> The superiority of the curved blade for cutting purposes is easily proved. In every cut the edge meets its object at some angle, and the penetrating portion becomes a wedge. But this wedge is not disposed at right angles with the Sword; the angle is more or less oblique according to the curvature and consequently it cuts with an acute edge.[12]

The European light cavalryman, for the most part, was armed with pistols, and a saber, as described above. Their role was to screen the movements of their army, keep in contact with the enemy, and cut the enemy's lines of communication. This role did not change for light cavalry for hundreds of years and was the role played by Napoleon's hussars and in America shown to brilliant effect by Confederate cavalry generals Stuart and Mosby, and later by the Union Army General Philip Sheridan.

Figure 6. Napoleonic cuirassiers in action as portrayed by the French military illustrator JOB, in the book *L'epopee du Costume Militaire Francais* by Henri Bouchot. Note the sword-hand position of the cuirassiers with the knuckle guard pointed up, the position still in use by the French cavalry when Patton studied at Saumur. (*Courtesy of the Biblioteca Virtual del Patrimonio Bibliographico*)

Although by the sixteenth century light and heavy cavalry carried weapons specific for their roles, light cavalry armed with sabers (for example, see Figure 3 above) and heavy cavalry with straight hand-and-half swords or more robust versions of what would eventually become civilian swords (see Figure 4 above), each soldier's weapon might be slightly different from that of his comrades. It wasn't until the mid-seventeenth century that armies began instituting the standardization of model weapons coinciding with the development of large standing armies. Sweden, Prussia, and France appear to be the earliest to institute regulation weapons in the middle of the seventeenth century, with other major powers following suit by the beginning of the next century.[13]

Over the period stretching from the Renaissance to the age of Napoleon, cavalry became more specialized and carried fewer types of weapons and eventually, for heavy cavalry, the sword moved from ancillary weapon to the primary weapon. Cuirassiers charged infantry squares, broadsword in hand.

Chapter 3

Cavalry Swords and Sabers in the Napoleonic Wars

The Revolution in France that brought down the Versailles court-centered world of Louis XVI created a republic hemmed in on all sides by enemy nations ruled by monarchs fearful of similar revolutions in their own countries and genuinely sympathetic to the plight of the royalty and nobility of France. Not surprisingly, the new French Republic created a revolutionary army. The French Revolution brought forth a national army and placed it on the world stage and other nations were forced to bring forth national armies on the same scale.[1] The Napoleonic Army that would emerge from the Revolution would be carefully studied and emulated, not least of all by the fledgling republic lately formed in North America – the United States.

Napoleon Bonaparte, with his political and military genius, directed the revolutionary ardor of the French and the nature of the revolutionary army toward the pursuits of finding the limits of French domination of Europe and the ultimate limit of glory on the battlefield. Clausewitz witnessed and understood the political and resulting military changes that occurred between the Ancien Régime and Napoleon and set it down in clear language for posterity:

> Bonaparte's audacity and luck have cast the old accepted practices to the winds. Major powers were shattered with virtually a single blow … wars waged by both sides to the full extent of their national strength must be conducted on different principles from wars in which policy was based on the comparative size of the regular armies. In those days, regular armies resembled navies, and were like them in their relation to the country and to its institutions. Fighting on land therefore had something in common with naval tactics, a quality which has now completely disappeared.[2]

In the Napoleonic age, as Clausewitz would explain, an army could conceivably be composed solely of cavalry, definitely, solely of infantry, but never solely of artillery. The effective army was at its most effective when it used all three branches. However, because of the improvements in firepower infantry became the most independent of all arms, and the least expensive – a battalion of 800 infantrymen cost the same to equip as a squadron of 150 cavalrymen or a battery of 8 field cannons.[3] But as the rifle and bayonet constituted the arms of the infantryman and the cannon the arms of the artilleryman, the sword was the primary weapon of the cavalryman, hence his prominence in this book.

A horse can carry about one-quarter of its weight. The average cuirassier and equipment weighed about 309lb, thus his horse would be about 1,240lb; a dragoon and his equipment weighed about 273lb so his horse would weigh about 1,100lb; and a light cavalryman like a hussar or chasseur and his equipment weighed about 251lb and so his horse would weigh about 1,000lb.[4] The biggest horses carried the biggest men with the heaviest equipment, cuirassiers with armor and heavy swords, the smaller horses, perhaps more nimble, carried light cavalry.

To sum up the role of the different types of cavalry under Napoleon: light cavalry formed a screen behind which the Grand Armée maneuvered, dragoons stiffened the screen, and heavy cavalry was held intact until the day of battle to smash and scatter the enemy.[5] For many armies, this description of the roles of cavalry would remain true for the next century. Although an artillery officer by training, Napoleon's universal genius sought improvements in all parts of the French Army – too many to be told here, but his improvements to the cavalry service are important

Figure 7. These French cuirassiers passing through Paris in 1914 have a similar appearance to their Napoleonic counterparts a century before.

in a book about swords. Napoleon improved efficiency by removing cavalry units tied to infantry divisions creating a cavalry reserve that Murat would command with distinction.[6]

From the time of Louis XIV to the time of Napoleon, the cavalry of an army amounted to anywhere from a quarter to a sixth of the number of infantrymen.[7] But Napoleon increased the effectiveness of the heavy cavalry by ordering that cuirassiers, and later the carabiniers, be armored not just with a breast plate but back plate, too. In addition they were issued steel helmets.[8]

Being completely armored had its advantages; at the Battle of Ratisbon, the French cuirassiers met their Austrian counterpoints and Marbot describes the encounter:

> Courage, tenacity and strength were well matched but the defensive arms were unequal, for the Austrian cuirasses only covered them in front and gave no protection to the back in a crowd. In this way the French troopers, who having double cuirasses and no fear of being wounded from behind, but only to think of thrusting, were able to give point at the enemy's back.[9]

Napoleon also improved the efficiency of the heavy cavalry by abolishing seven regiments to improve the strength of the remaining eighteen regiments by putting the largest troopers and tallest horses into the first twelve and converting the remainder into the dragoons, who, although they bore heavy straight swords, were generally smaller and rode smaller horses. Simultaneously, he converted the three newest hussar regiments to dragoons, much to the chagrin of the hussar officers who had to re-equip themselves with straight swords instead of their curved sabers and larger horses, and buy new uniforms, all at extra expense.[10]

The heavy cavalry that Napoleon inherited were armed with a mixture of swords, French and foreign, but early in his reign, a long, straight-bladed model was adopted for all cuirassiers. The 1803 Model had a blade 38in long – a big sword that was difficult to master and thus not always popular with the troopers who carried them (see Figure 8 opposite).[11]

A logical question might be, why did heavy cavalry carry a long straight sword and light cavalry a curved saber. One reason might be that the long straight sword could be used like a lance, but had the advantages of being able to be used as a cut-and-thrust sword after the charge had terminated and the cuirassier found himself in a melee; with a sword, he could parry the attacks of his opponents. A lance is a great thrusting weapon, but virtually useless once the charge is over. If the charge does not carry the trooper clear of the enemy lines, he would find himself in a melee and would have to be able to attack and defend – to cut and parry.

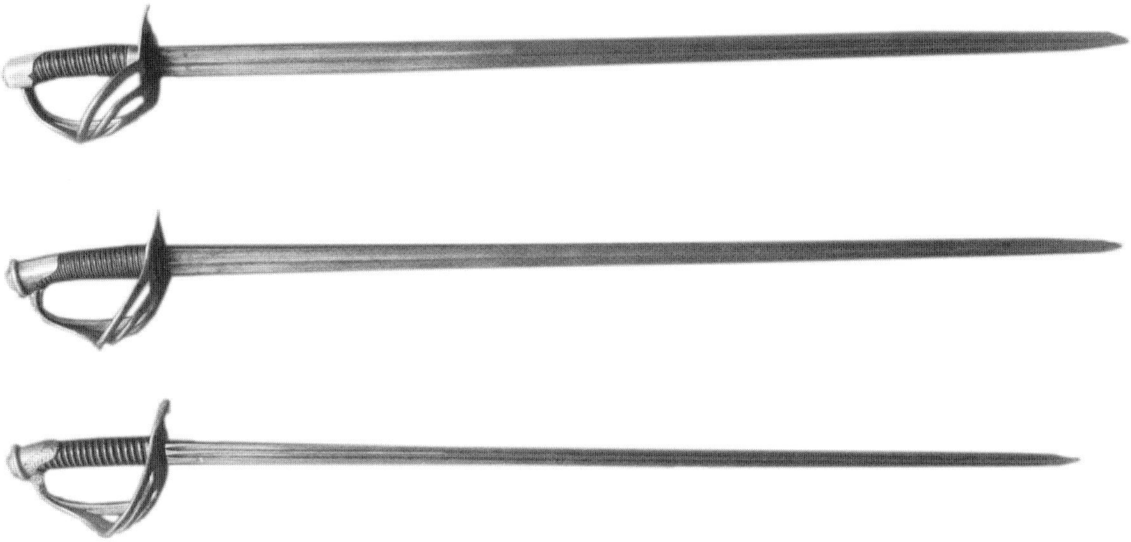

Figure 8. French cuirassier broadswords, from top to bottom: Model 1803: Marked 1814; Model 1816: Marked 1834. (*Graham Collection*); and Model 1850 Officers' Sword. (*Fogle Collection)*

The cuirassier's broadsword was long enough to provide the thrust like a lance; however, it could be used after the charge had terminated to parry attacks and to riposte with the point or cut with the edge.

Light cavalry had a different role and carried lighter curved sabers, better for cutting but not as well suited to thrusting. As Figures 3 (above) and 11 (below) illustrate, light cavalry sabers changed little from the seventeenth century to the beginning of the nineteenth century. These examples are marked by light hand protection and robust, significantly curved single-edged blades. (For the most part, mounted soldiers in the newly formed United States would adopt rugged curved sabers similar to European counterparts, if often a bit cruder, see Figure 25 below.)

In the Napoleonic light cavalry, there was a greater variety of saber types, most had curved blades between 32 and 36in, good for cutting, but capable of giving point. Most had simple stirrup guards, but some had guards that afforded greater hand protection, with additional bars beside the knuckle guard.

As to why the saber was the more popular type of sword to be carried by cavalry, perhaps the best answer is given in *The Book of the Sword* by the Victorian soldier and explorer Sir Richard Burton, not to be confused with the twentieth-century movie star of the same name. Burton was an accomplished fencer and needed to be, more than once he defended his life with the sword. About those who fight with swords from horseback he wrote:

> [They] preferred for the best of reasons the curved type. The straight Sword, used only for thrusting, is hard to handle when the horse moves swiftly; and

Figure 9. A French light cavalry trooper's saber from the age of Napoleon. (*Fogle Collection*)

> the broad straight blade loses its value by the length of the plane along which it has to travel. On the other hand, the bent blade collects, like the battle-axe, all the momentum at the 'half-weak,' or centre of percussion, where the curve is greatest. Lastly the 'drawing-cut' would be easier to the mounted man and would most injure his enemy.[12]

Many good examples of the life of the light cavalry in the age of Napoleon are provided by the Baron de Marbot in his memoirs, which detail his service under Napoleon from his enlistment into the First Hussars in 1799 at age 16 until the Emperor's defeat at Leipzig. His first two actions illustrate the role of the light cavalry.

In his first action, Marbot and some specially selected men of the regiment, by stealth, crossed into the hostile territory of the Austrians in what is now northern Italy. Near the commune of Dego, they acquired by force the cattle belonging to their rival cavalry unit on the Austrian side and brought the much-needed provisions to their own camp without firing a shot. In his second action, Marbot has command thrust upon him and with it the task of finding contact with the enemy for the infantry division of General Seras, to whom his company is attached. Marbot interrogates locals, finds the enemy's cavalry camp, turns the position to his favor, and catches the Austrian cavalry while watering their horses. His company captures more than a dozen of the enemy and brings them and their intelligence to his general. Neither of these suggest the glory of a cavalry charge at the height of battle, but, rather the day-to-day work of foraging for provisions and screening the movements of the army while providing intelligence about the enemy – the work of light cavalry.[13]

Marbot's light cavalry career during the Napoleonic Wars began as a trooper and ended as the colonel of two regiments, first the 23rd Chasseurs au Cheval,

Figure 10. Some light cavalry favored a less curved light-cavalry saber. (*Fogle Collection*)

and then the Seventh Hussars. He commanded the 23rd Chasseurs during the Russian campaign, and provides one of many informative descriptions about the use of the saber in war. Outside of Wilkomir, Ukmerge in present-day Lithuania, Marbot charged his regiment across a bridge against a wall of Russian grenadiers, the following quote describes the result: 'Having pierced the enemy's ranks, my chasseurs, dexterously using their points, did fearful execution.'[14] Light cavalry could give point with their curved sabers as well as cut.

Soon after Wilkomir, the Russians marched on the post-house of Kliastitsi where the right of the French Army rested, and Marbot describes the cavalry action in which he took part:

> As soon as the 23rd were formed in line we marched upon the Russian infantry, which halted and steadily awaited us; it was the Tamboff regiment. When we were within striking distance I gave the word to charge. This was carried out all the more efficiently for the stimulus which the fact that their comrades of the 24th were watching them gave to my troopers. The enemy committed the serious blunder, as I think it is, of spending all his fire at once, by giving us a volley, which badly aimed as it was emptied but a few saddles. A file fire would have been far more destructive. Before the Russians could reload we were upon them at full speed of our excellent horses, and the shock was so violent that they were overthrown in heaps. Many rose again and tried to defend themselves with the bayonet against the troopers' points; but after losing heavily they fell back, and at last broke.[15]

The charge against the Tamboff Regiment shows that during the Napoleonic Wars, even light cavalry could still effectively charge against infantry in position; a generation later, cavalry in the American Civil War could rarely succeed in charging against infantry armed with rifled muskets.

The two greatest cavalry generals under Napoleon were Marshal Murat and General Lasalle. Antoine Charles Lasalle entered the French Revolutionary Army and caught Napoleon's attention at the Battle of Rivoli, when the Austrian infantry tried to escape down a rocky escarpment to the valley below; although warned that pursuit on horseback was impossible, Lasalle led his men down the steep rocks and captured a thousand prisoners. After the French victory at the Battle of Jena, Lasalle showed what a regiment of hussars could accomplish. By ruse, he convinced the governor of the fortified city of Stettin (now Szczecin, Poland) to surrender.[16] In this action, with only 900 hussars, Lasalle captured 5,000 prisoners and 200 guns.[17]

When the great cavalry commanders are named, Joachim Murat is often foremost. He was Napoleon's cavalry commander from 1800 through 1807.[18] Murat distinguished himself during the French victory at Eylau, when he led a cavalry charge of ninety squadrons; 'the terrible weight of this mass broke the Russian center, upon which it charged with the sabre, and threw it into complete disorder.'[19]

In America, the Civil War cavalry commander J.E.B. Stuart most closely resembled the bold leaders of horsemen of Napoleonic fame, as will be shown later.

In his epic description of Napoleon's Grande Armée, the author Colonel John R. Elting provides thorough descriptions of all of Napoleon's marshals; introducing them, he names only one American general of the Second World War who would have been at home among them, George S. Patton, who thoroughly understood the role of cavalry and adopted the abilities of horse cavalry for modern mechanized units.[20]

Perhaps the role of cavalry was best described in 1916 by Patton himself:

> The first duty of cavalry is to discover the movements of the chief columns of the enemy, and at the same time to prevent the hostile cavalry from learning the whereabouts of our main body. To accomplish this, its dual mission, it must defeat the hostile cavalry and must do it quickly. Quick action means the saber![21]

Chapter 4

From Curved Sabers to Straight Cavalry Swords, or One Type of Sword for All Cavalry

The 400-year question of whether straight or curved swords were better for cavalry would be answered during the late nineteenth and early twentieth centuries, in favor of straight cavalry swords. Over the course of the nineteenth century, most European cavalry units began to be equipped with one kind of sword light cavalry sabers; the heavy cavalry swords and gleaming cuirasses becoming the kind of equipment relegated to royal bodyguards.

In 1821, the British Army adopted a new light cavalry saber. The new weapon abandoned generations of severely curved cutting weapons, culminating in the 1796 light cavalry saber, perhaps the finest mass-produced cutting weapon ever made (see Figure 11 below). The 1821 light cavalry saber, however, was a departure and a return. Gone was the 'P' knuckle bow of the previous model, replaced with a three-

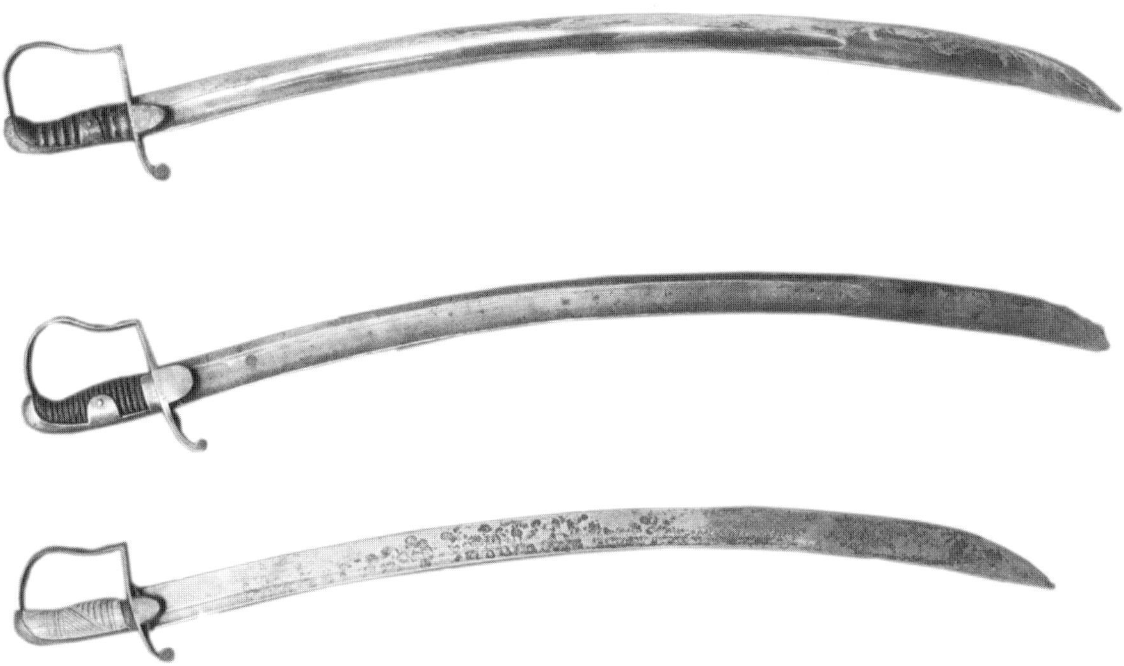

Figure 11. Light cavalry sabers, from top to bottom: British Model 1796 Light Cavalry Saber; Prussian Model 1811 'Blucher' Light Cavalry Saber; US Saber c. 1830. (*Fogle Collection*)

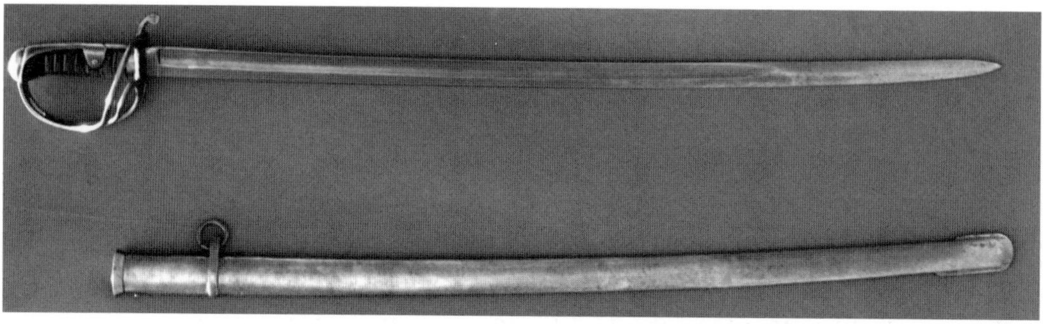

Figure 12. British Model 1821 Light Cavalry Saber, with a blade made in Birmingham, England, probably a Confederate purchase. The author purchased this example in Atlanta in the late 1980s. The Model 1821 which was almost straight and with a spear point was a departure from the heavily curved cutting sword that was its predecessor. (*Author's Collection*)

bar guard providing much greater protection to the sword hand; also gone was the severely curved blade, good only for cutting, replaced with an almost straight single-edged blade with a false edge extending several inches from the point. Those are the changes from the previous light cavalry sabers, but in a way it was a return to an earlier sword, the multi-purpose cut-and-thrust sword popular from the time of Cromwell. The blade of the 1821 sword would continue with minor changes through many models, with the primary change being improvements to the hand protection.

Many Model 1821 British light cavalry sabers were shipped to both the north and the south during the American Civil War. The famous Tiffany swords purchased during the Civil War bear a great resemblance to the British Model 1821 Light Cavalry Saber.

In Britain, most heavy and light cavalry regiments began using the new pattern 1853 sword which was similar to the 1821.[1] One difference with the new pattern was that it had a kitchen knife-like tang and a handle attached to the handle scales

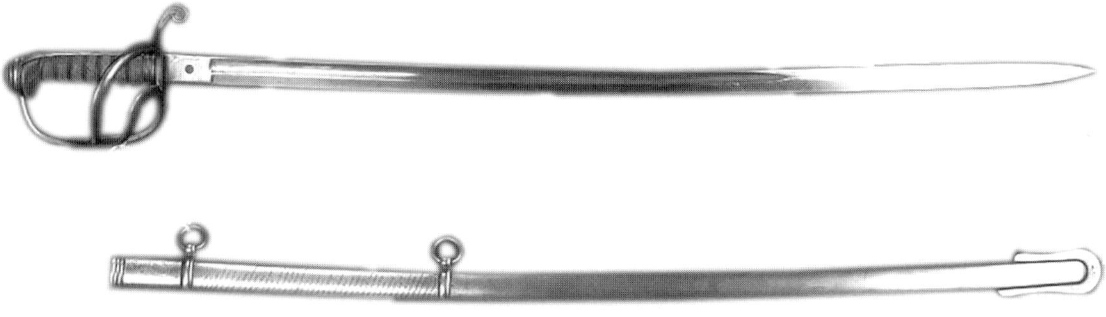

Figure 13. British Model 1821 Light Cavalry Saber carried by Union Major General Frank Wheaton. (*Courtesy of the Smithsonian Institution, National Museum of American History*)

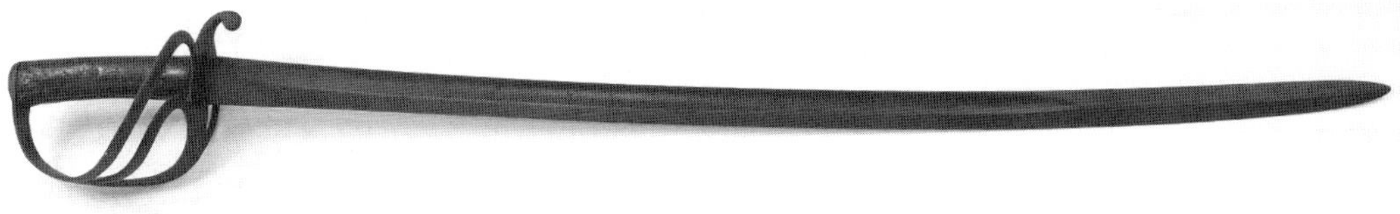

Figure 14. British Model 1853 Cavalry Trooper's Saber. Similar to the Model 1821, it was the first British saber for all cavalry, heavy and light. Unlike previous sabers where the tang was attached to the handle, with the model 1853, the blade continued as a flat piece of steel to the pommel with handle scales riveted to each side like a kitchen knife for greater durability. Sixty years later, the US would adopt a similar approach with its Model 1913 Cavalry Saber. (*Courtesy of Easton Antique Arms*)

to the blade with rivets. In 1864, Britain would replace the three-bar guard with a solid sheet of steel perforated with a Maltese cross cut-out. All of the light cavalry sabers of the British cavalry would remain, for the most part, slightly curved cut-and-thrust sabers, the blades differing little from the Model 1821.

In 1894, almost a decade after Patton was born, Sir Richard Burton's *The Book of the Sword* was published. In his book, Burton describes the history of the sword and also how it was used in his own day. Although, as seen previously, he explained why cutting swords were popular with cavalry, he advocated the thrust over the cut. 'Indeed, the man who first gave point made a discovery which more than doubled the capability of his weapon.'[2] Further, Burton stated, 'there is no question of the superiority between the thrust and the cut … who delivers point has an advantage in time and distance over [the one] who uses [the] edge.'[3]

Burton further argued that puncture wounds were more often fatal than cuts, and supports his statement with the findings of hospitals and examples from military history – Napoleon's exhortation to his cavalry 'to give point' at the Battle of Aspronne. Patton would make similar arguments in favor of a thrust-centric cavalry sword twenty years later. The British, for their part, would adopt a straight

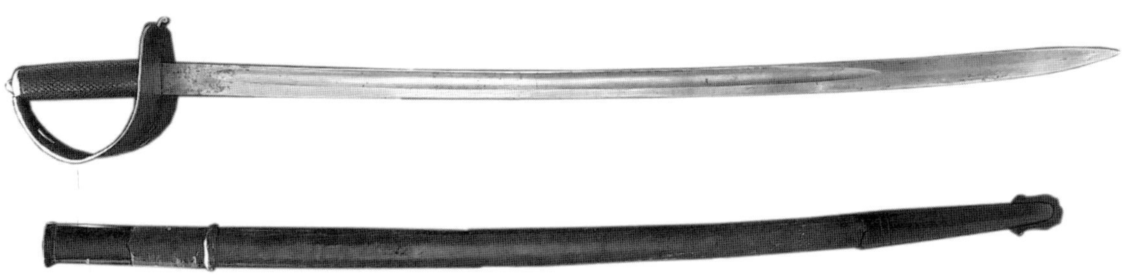

Figure 15. British Model 1864 Cavalry Saber. (*Graham Collection*)

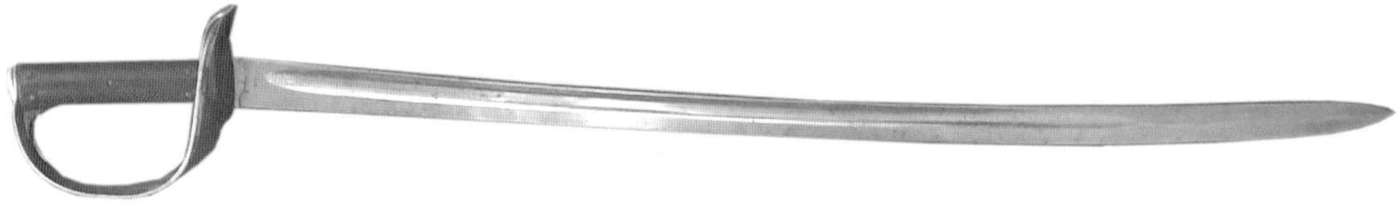

Figure 16. Model 1882 British Cavalry Troopers' Saber. (*Graham Collection*)

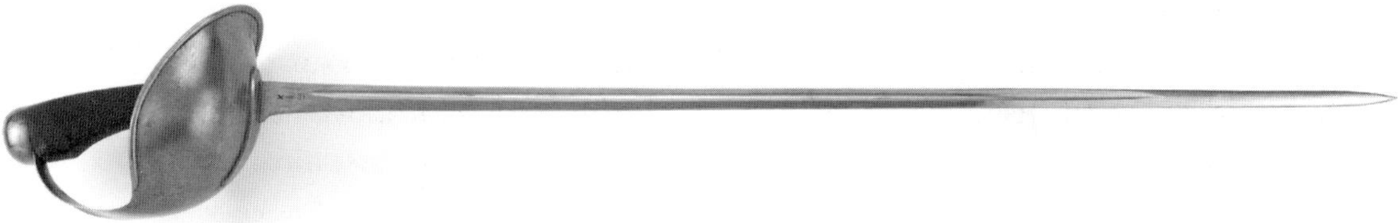

Figure 17. British Cavalry Saber Model 1908. Like the Patton saber that followed it, it is a straight, thrusting cavalry sword with a sheet-steel bowl guard for good hand protection, but the blade, with its 'T' cross section, precluded cutting. However, the Patton saber is double-edged making it a good cutting sword and its grip is in line with the blade. (*Courtesy of the Springfield Armory Museum*)

cavalry sword, the Model 1908. With its 'T' cross-section, it basically does not allow for the cut and is purely a thrusting weapon, rather like a skewer.

By the end of the nineteenth century, the French would adopt a straight sword for their cavalry troopers, finally replacing the Model 1822 (see Figures 18 and 31 below). The Model 1882, while retaining the brass guard of French swords going back to the Napoleonic era and before, had a stiff, straight blade, ideal for the thrust, but with enough edge and blade geometry to cut. In turn, the French would replace the 1882 in 1896 with another straight sword, but with a symmetrical guard.

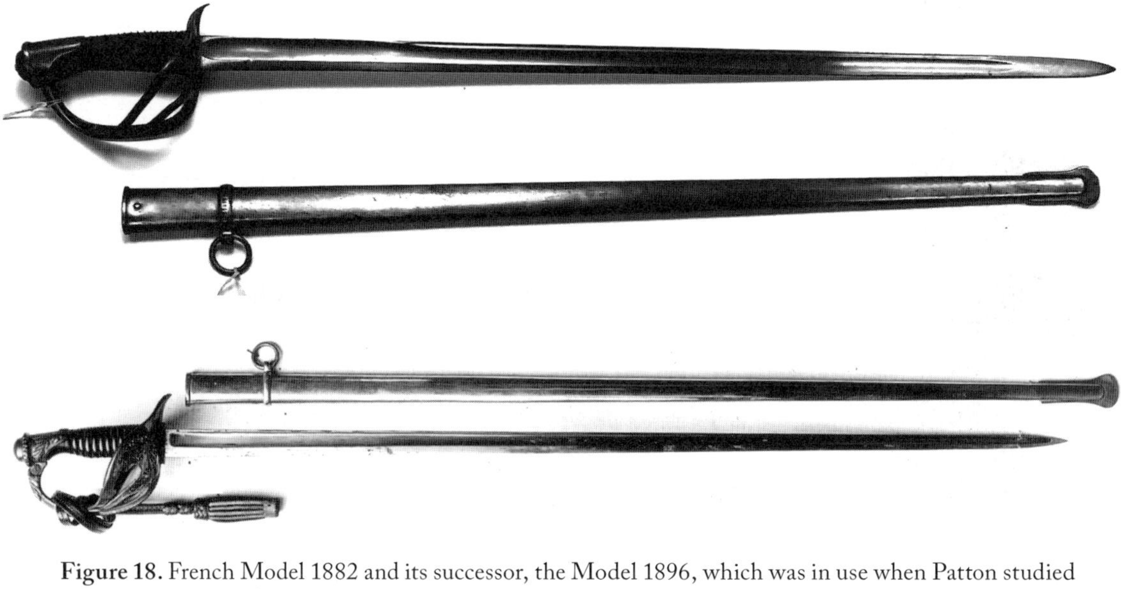

Figure 18. French Model 1882 and its successor, the Model 1896, which was in use when Patton studied fencing at Saumur. (*Courtesy of the Smithsonian Institution, National Museum of American History*)

The Evolution of Sword Design in Germany

The most influential military power among the German states during the eighteenth and nineteenth centuries was Prussia. During the late eighteenth century, the Prussians were provided with some military materiel by their ally, Great Britain, in the struggle against Napoleon, including British cavalry sabers. The Prussians so liked the Model 1796 British Light Cavalry Sabers they received that they began making their own version – the Model 1811 Cavalry Saber, often referred to as the Blucher saber, so called after Prussia's most important field marshal against Napoleon. Cavalry units in the United States would adopt swords similar to the British Model 1796 and the Blucher saber during the early years of the republic. (See Figure 11 above)

The Model 1811 would remain the Prussian cavalry saber until the late 1840s. Prussian light cavalry sabers illustrate an interesting trend which would be followed by many Western European nations. During the Napoleonic Wars, the Prussians favored significantly curved, hatched pointed sabers, but over time, the hatchet point was replaced with a spear point, better for thrusting. By the 1850s greater hand protection was added and the sword became straighter, again improving it for thrusting. By 1889, the cavalry saber was replaced with a cavalry sword – straight, quill-backed, and with an ergonomic hilt, ideal for thrusting.

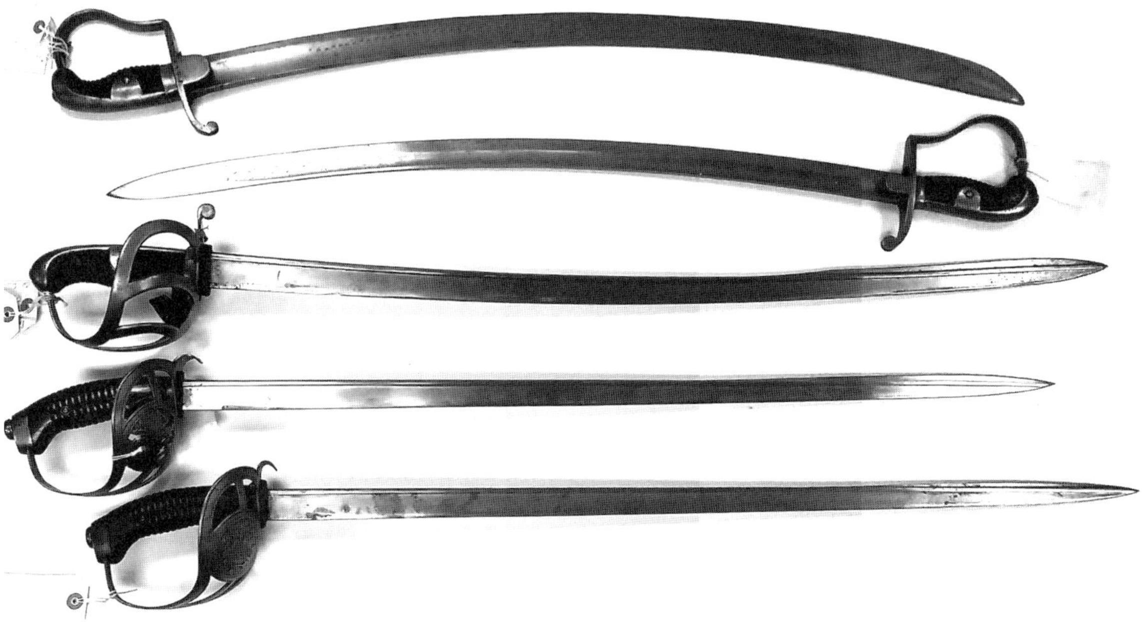

Figure 19. Prussian sabers, from top to bottom: Model 1811, 1838, 1852–79, and two examples of Model 1889 – the top is Prussian, the bottom an export to the Chinese Imperial Army. (*Courtesy of the Smithsonian Institution, National Museum of American History*)

In 1893, Sweden replaced the robust single-edged cutting sword, which it had used for fifty years, with an equally robust straight double-edged cavalry sword, designed for thrusting. It is important to note that Sweden, the largest of the Scandinavian countries, had a long, proud cavalry tradition. As Swedish cavalry officers won all the medals in the modern pentathlon events of the 1912 Olympic Games, Patton who competed against them in those games was probably impressed by their cavalry swords. The similarity between the Swedish cavalry sword and the

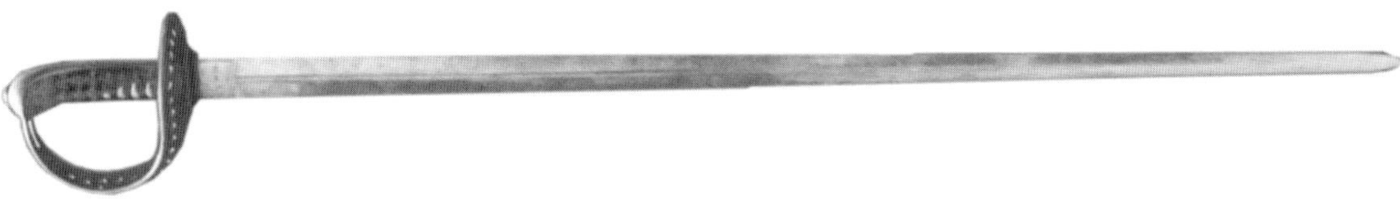

Figure 20. Swedish Model 1893 Cavalry Saber. Like the Patton saber, it is double-edged and the grip is in line with the blade. (*Graham Collection*)

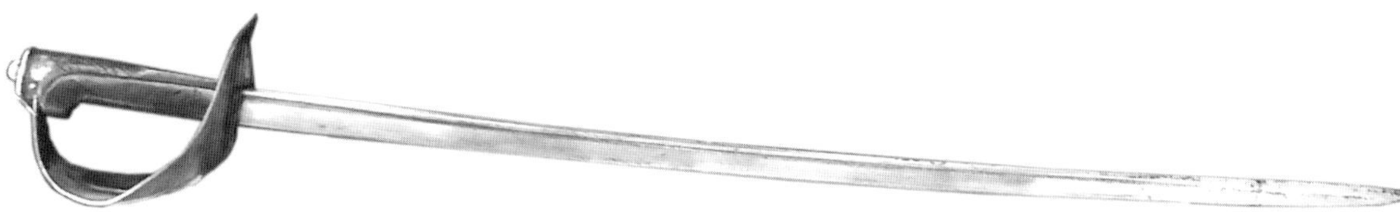

Figure 21. Italian Cavalry Troopers' Saber Model 1873. Note that the blade is only slightly curved and quill-backed, for better stiffness. The guard is sheet-steel, with a depression for the thumb in the guard, making it an effective thrusting weapon. (*Fogle Collection*)

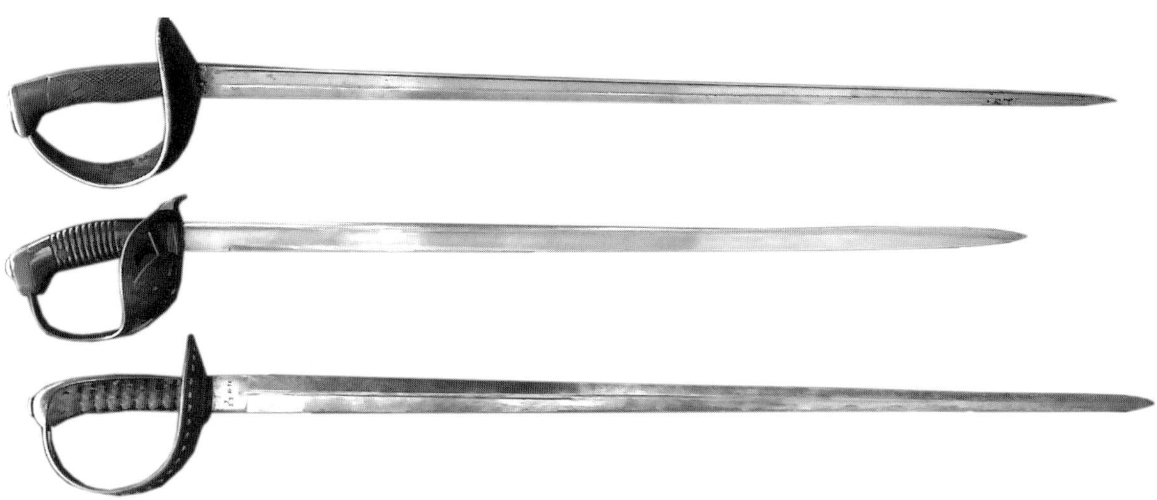

Figure 22. Comparison of late nineteenth- and early twentieth-century cavalry swords; from top to bottom: Spanish Model 1907 Cavalry Saber. (*Fogle Collection*); Prussian Model 1889 Cavalry Saber; and the Swedish Cavalry Saber Model 1893. (*Graham Collection*); the Swedish sword may be the closest precursor of the sword designed by Patton.

cavalry sword Patton developed is striking; both are straight double-edged swords, with hilts in line with their long blades. In 1889, Prussia replaced its light cavalry saber with a light cavalry sword – straight and pipe-backed (see Figure 19 above). Spain would adopt a straight cavalry sword in 1907. Italy had adopted a straight cavalry saber by 1873, coinciding with the development of the modern fencing saber. As described earlier, Great Britain would follow suit in 1908 with a straight thrusting sword with a blade that had a 'T' cross-section that virtually precluded cutting (see Figure 17 above).

The ordnance departments of European nations required not only straight blades, good for thrusting as well as cutting, for their cavalry troopers, but also that they be equipped with swords offering better hand protection. Gone were the multi-barred hilts, replaced with solid guards of sheet metal to better protect the hand. Patton would take into consideration all of these developments when he designed the army's last cavalry sword.

Chapter 5

A Survey of Swords in Colonial America and the US Army

The Bowie Knife and the Tomahawk are uniquely American weapons. Although the steel knife was brought to the western hemisphere by European explorers centuries earlier, Jim Bowie's improvement was universally acclaimed and is still popular today. The Tomahawk favored by native Americans, in its best-known form, had a relatively short wooden handle with an iron or steel hatchet blade, often backed by a pipe bowl, and that form was only possible because of the trade between native Americans and colonists and explorers from Europe.

Similarly, firearms were brought to the Americas by explorers and colonists, however, their descendants would make such improvements that their firearms would forever carry their names and indeed be synonymous with the names of their creators – Colt and Winchester being the foremost examples. Samuel Colt did not invent the revolver, but his improvement which turned and locked the cylinder with each draw of the hammer made his revolver a formidable and practical weapon. O.F. Winchester did not invent the repeating rifle, but by improving the Henry rifle, his name became synonymous with lever-action rifles.

Swords of the Colonists

Swords were important weapons to the early colonists and explorers, and were part of the inventory of US ordnance up until the Second World War. Next to firearms, the most important weapon that European explorers and colonists brought to the Americas was the sword, and this was particularly true for the early colonists who needed a reliable weapon for close fighting once they had discharged their muskets.

The Pequot War

Perhaps the most significant of the early struggles undertaken by the English settlers of New England was against the Pequot Tribe in what is now Connecticut. The origin of the Pequot War was the settlement of English colonists, who had formerly lived in Watertown, Dorchester, and Newton, resettling in the Connecticut Valley in 1636. Attacks by the Pequots against the settlers resulted in the organization under Captain John Mason of a force to make war against the Pequots. Beginning on May 10, 1637, Captain Mason with ninety men under his command and Captain Underhill with another twenty men sailed to Narragansett Bay and then marched to the Pequot fort near the Mystic River.

On the brow of a hill stood the Pequot fort which comprised a circular stockade, 12ft high, with two entrances on opposing sides – inside were many wigwams.[1]

Mason led his assault by the light of a bright moon, but their advance was given away by a barking dog and taken up by a sentinel who shouted the warning 'Owanux! Owanux!' Mason and his men fired a volley from their muskets and then charged the fort, sword in hand, Mason to one entrance and Underhill to the other. In the confined fort, there was no room for loading and aiming muskets, and the English fought the Pequots hand-to-hand, hewing them with their swords. Mason seized a firebrand and set one of the wigwams alight and the fire quickly spread.[2]

The attack on the fort was a complete success for the English colonists, for whom the sword was an important weapon. Mason's loyal sergeant William Heydon cut the bowstring of a bow with which a Pequot was trying to shoot Captain Mason.[3] The sword is in the collection of the Connecticut Historical Society (see Figure 23 below).

Although the destruction of the Mystic River fort of the Pequots did not end the war that night, the power of the Pequots was broken.

Figure 23. The sword of Sergeant William Hayden which was carried by Hayden during the Pequot Wars and with which he saved the life of Captain Mason. (*Courtesy of the Connecticut Historical Society*)

War for Independence

When the thirteen colonies in North America united in their cause for independence from British rule the secession took the form of armed revolt with the colonies organizing militias and forming the Continental Army.

Looking at the Revolutionary War, one can well appreciate the observation by Clausewitz that armies composed solely of infantry are perfectly able to operate independently and that armies of infantry supported by artillery can win wars.

Although the Continental Army had some cavalry, the war contained no significant cavalry charges and its outcome, for the most part, did not turn on the success of that arm.

Although the Continental Congress legislated the raising of four regiments of light dragoons late in 1776, no such units were actually raised until the following year. During 1777 and 1778, Baylor's Dragoons provided armed escort for General George Washington, and served as patrols, videttes, and couriers for his headquarters. They were uniformed in white coats faced with light blue, buff trousers, and high boots. They were armed with single-shot flintlock pistols, carbines when they could get them, and heavy cavalry sabers.[4]

Although much of the struggle for independence was between infantry units, there was considerable cavalry activity in South Carolina, where Francis Marion, Henry (Lighthorse Harry) Lee, and William Washinigton fought for the liberty of the thirteen colonies against Banastre Tarlton and his Green Dragoons.

Many American Revolutionary cavalry leaders considered swords to be the only important weapon carried by cavalry men and famous leaders such as Henry Lee and William Washington disdained cavalry use of firearms.[5]

As an example of the importance of the sword as a cavalry weapon, Francis Marion converted some of his infantry to cavalry because of a shortage of powder and shot. He then procured all of the whipsaws in his area of Whig support and had blacksmiths convert them to broadswords for his former infantry, then newly mounted as cavalry. The teaching that cavalry armed with swords will overcome cavalry armed with firearms was held as a maxim of military thought from the War for Independence through the Civil War.[6]

Figure 24. Revolutionary War cavalry saber. (*Courtesy of David Condon Antique Arms, Middleburg, Virginia*)

Swords of the New Republic

Throughout the history of the US Army, its swords were modeled on and in some cases replicas of European swords. Some of the earliest US cavalry swords – the Rose Contract 1803, Horseman 1807, Starr 1808 and 1812 and 1818, among others – resembled the British and Prussian light cavalry sabers (see Figure 11 above).[7] Nathan Starr was the nation's foremost manufacturer of swords, but in 1828, he

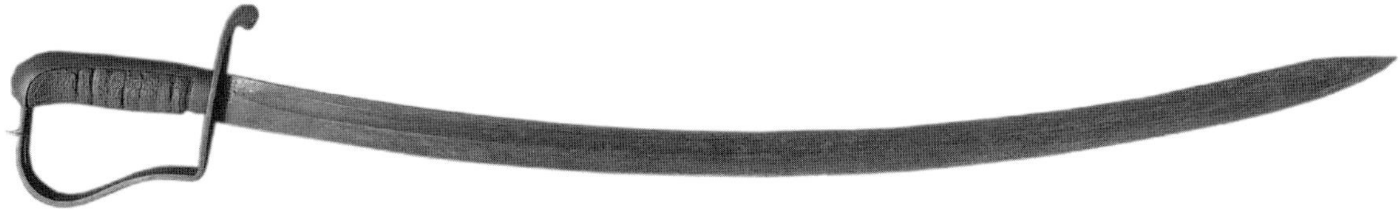

Figure 25. Starr Cavalry Saber Model 1818. (*Graham Collection*)

decided to concentrate his company's energies solely on designing and manufacturing firearms leaving the field of edged weapons open to the Ames family. N.P. Ames would establish his factory in Chicopee, Massachusetts, taking advantage of the hydropower provided by the Chicopee River. Ames would become the most productive sword manufacturer in the United States during the nineteenth century.[8]

In 1833, the United States established a regiment of dragoons, a word that originally suggested mounted infantry, but by the nineteenth century was just another word for heavy cavalry. The sword developed for the dragoons, the Model 1833, was manufactured by the Ames Sword Company and resembled British quill-back swords of the period.[9] However, hard use in the field caused dragoons to complain of bent blades and scabbards.[10] Nonetheless, the Model 1833 would continue in somewhat limited use with state militias even up until the Civil War. Additionally, the pendulum was swinging away from British influence over Americans in military science; soon the United States would look to France for an example of military excellence.

Figure 26. Cabotville, Massachusetts, at the time Ames manufactured the Model 1833 Dragoon Saber and other swords for the military. Cabotville became part of the newly incorporated town of Chicopee.

Coincidentally, a year before the US Army adopted the Dragoon Saber, it adopted a sword for general officers and their staffs, the General and Staff Officers' Sword, 1832.[11] The sword was an American copy of the British Model 1796 dress sword for officers of heavy cavalry.[12] Although it has a hilt based on dueling swords of the period, it has a robust straight double-edged blade, and except that its blade is about 3in shorter, it is more similar to the blade on the cavalry sword developed by George S. Patton than any other sword adopted by the US Army.

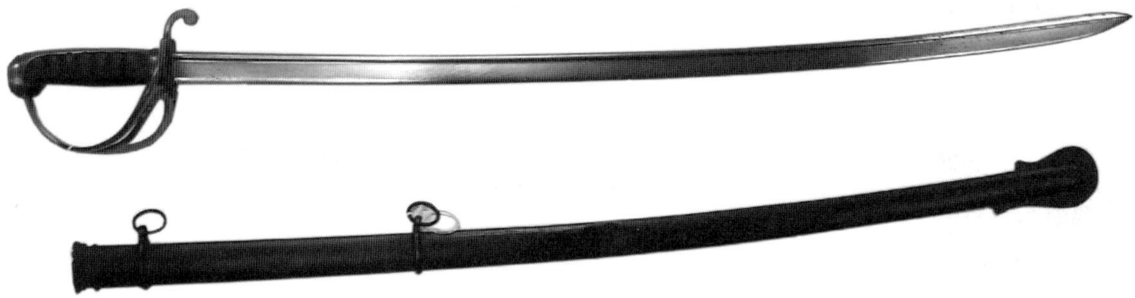

Figure 27. US Model 1833 Dragoon Saber and Scabbard. (*Courtesy of the Smithsonian Institution, National Museum of American History*)

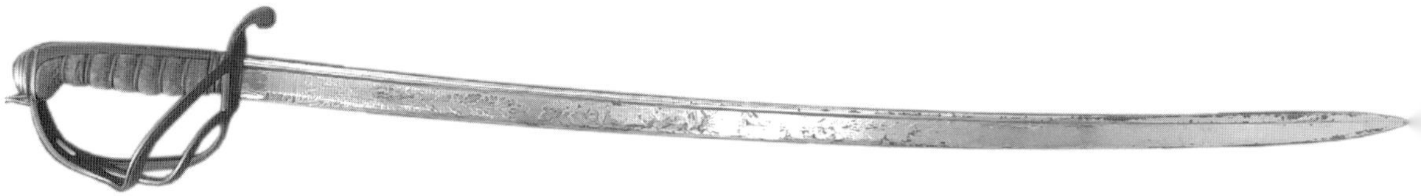

Figure 28. 1860 Connecticut State copy of the 1833 Dragoon, made by Horstman for Union officers in the Civil War. (*Fogle Collection*)

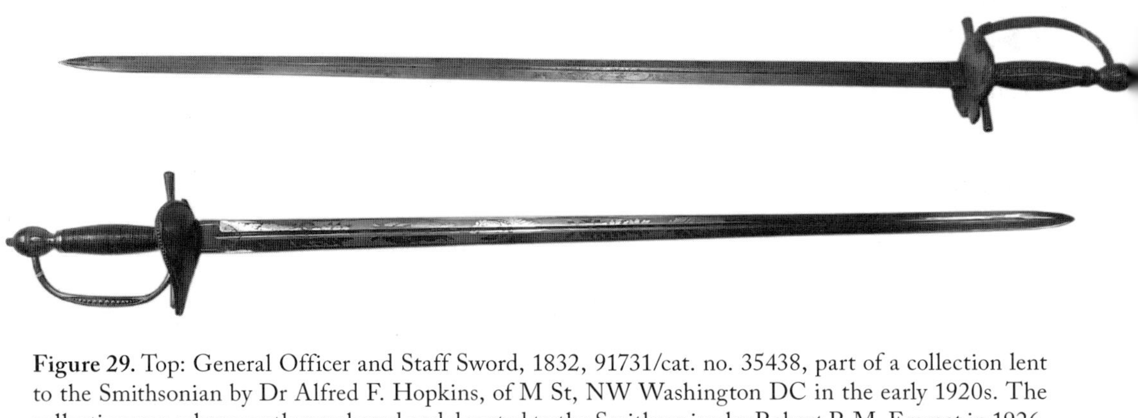

Figure 29. Top: General Officer and Staff Sword, 1832, 91731/cat. no. 35438, part of a collection lent to the Smithsonian by Dr Alfred F. Hopkins, of M St, NW Washington DC in the early 1920s. The collection was subsequently purchased and donated to the Smithsonian by Robert R.M. Emmet in 1926. Bottom: A.36367 c.8044, owned by Thomas Swords, apparently a West Point graduate, commissioned a 2nd Lieutenant of Infantry in 1832 and subsequently served in the Quartermaster Corps. He retired a Brevet Major General. The sword was donated in 1900 by his sister-in-law. (*Courtesy of the Smithsonian Institution, National Museum of American History*)

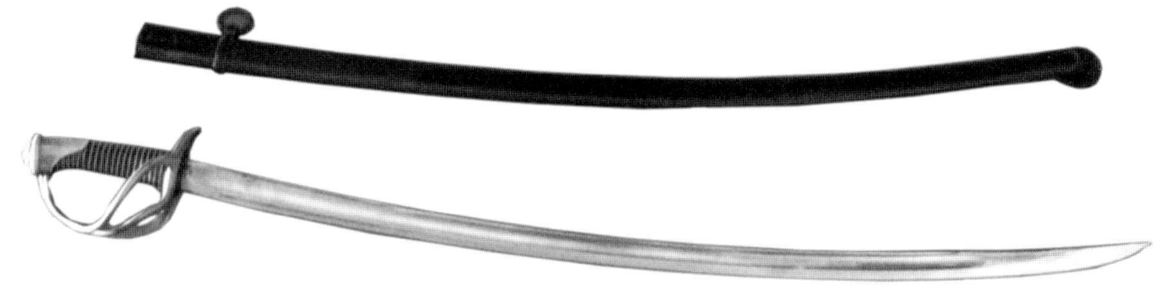

Figure 30. French Model 1822 Cavalry Saber, manufactured in 1879. (*Courtesy of the Smithsonian Institution, National Museum of American History*)

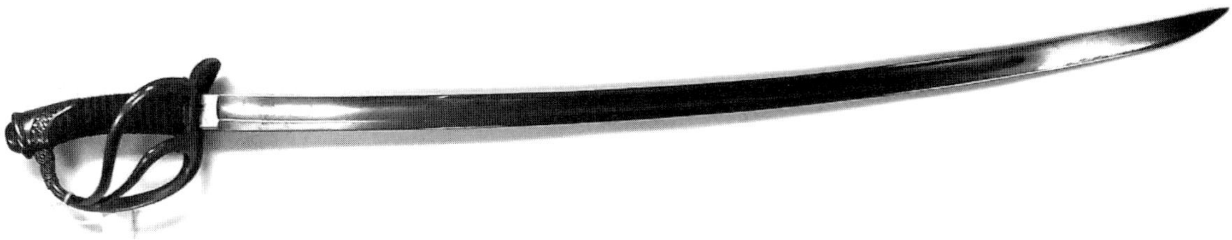

Figure 31. US Saber Model 1840, the type carried by cavalry officers serving in Mexico. (*Courtesy of the Smithsonian Institution, National Museum of American History*)

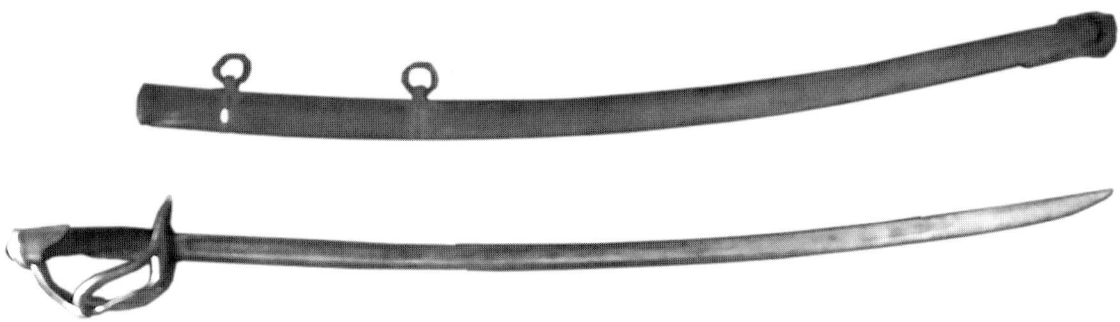

Figure 32. French manufactured US Model 1840 Dragoon Saber. (*Graham Collection*)

French Influence

From the standpoint of military history, the first half of the nineteenth century is dominated by the campaigns of Napoleon, and even after his personal power and prominence came to an end with his defeat at Waterloo in June 1815, his battles and strategy were more closely studied than those of the generals who defeated him. In addition, his grand-nephew, Napoleon III, continued to enhance French military prestige. During his reign, the French developed one of the most popular field pieces used during the American Civil War, the smoothbore 12-pounder, a cannon commonly referred to as a 'Napoleon,' after Napoleon III.

Further, much that was superficial about the French Army was copied, and the United States was among the many imitators. Among the items purchased from France were swords, and America also copied French sword designs. In 1840 the US Army replaced the 1833 dragoon saber with a copy of the French Model 1822 Cavalry Saber (see Figures 31 and 32 above).

It had a brass guard with three branches, including the knuckle bow and an almost 35in single-edged steel blade, with two fullers, one narrow and one shallow and broad.[13] It is important to note that this sword, the Model 1822 Cavalry Troopers' Saber, is carried today by the ceremonial cavalry troopers of the French Republican Guard.

In some cases, the US Model 1840 Dragoon Saber was not just an American reproduction of the French cavalry trooper's sword, there are examples that were manufactured in France (see Figure 32 above). Like the French saber, the 1840 Dragoon was a robust slightly curved saber with good hand protection and considerable weight, leading to it gaining the nickname 'the old wrist breaker.'

Figure 33. Department of the Army Poster No. 21–40: 'Remember Your Regiment and Follow Your Officers!'. It shows Captain Charles Augustus May saber charging the Mexican artillery at the Battle of Resaca de la Palma.

The Mexican-American War

President James K. Polk's inauguration speech in March 1845 was so bellicose regarding his determination for the United States to annex Texas, that after he concluded, the Mexican Minister in Washington demanded his passports and officially broke diplomatic relations between Mexico and the United States.[14] There was little doubt that there would be war between the two countries, which the Polk administration began planning almost immediately. In June 1845, Texans voted overwhelmingly to join the Union, and shortly after, Zachary Taylor moved his command of about 3,000 men to Corpus Christi on the Nueces River.

Texas at the time was a Mexican state with a southern border on the Nueces River, about 160 miles north of the present border on the Rio Grande. On January 13, the War Department ordered Taylor to move his command from the Nueces River to the north bank of the Rio Grande, then still Mexican territory, and on March 8, Taylor and his command began their overland march.

The US cavalry was composed of seven companies of the Second US Dragoons, under the command of Colonel Twiggs, which were armed with Colt revolvers and Model 1840 Dragoon Sabers like the one seen in Figure 31 above. The cavalry was followed by three brigades of infantry, one of the junior infantry officers was Ulysses S. Grant who detailed his experiences in Mexico in his celebrated memoirs. In mid-March, the army reached the Rio Grande, opposite Matamoros, and immediately began fortifying the position.[15] When the work was completed, the command under Zachary Taylor marched to Point Isabel for badly needed supplies. While Taylor was away, the garrison in the fortified position, called Fort Texas, was attacked by the Mexican Army, and the Mexican-American War had begun, roughly ten years after the Battle of the Alamo. The Mexican commander, General Mariano Arista, realized that he did not have enough artillery to capture the fort and further he had heard that the Americans were on the move from Point Isabel.

The armies met not far from Point Isabel at Palo Alto, so called on account of the tall trees that stood out from the chest-high grass. It was primarily an infantry and artillery engagement where the infantry and artillery of the US Army showed themselves to be superior to those branches of the Mexican Army, especially the artillery. The US artillery not only had more cannons, but these were generally of bigger caliber and with better ammunition.[16] Although the US Army was outnumbered, it held its own and at the end of the day a truce was called; in the morning the Mexicans withdrew to find a more defensible position – a gulch called by the North Americans Resaca de la Palma. Amidst the dense chaparral and a few palm trees, General Arista formed his line on the south side of a dry riverbed that ran roughly east to west.

Remember Your Regiment and Follow Your Officers!

Taylor ordered 150 infantry under Captains C.F. Smith and McCall to perform a reconnaissance in force.[17] The infantry scouts followed the Matamoros Road to Resaca de la Palma and found Arista's position, and after they reported, Taylor devised a simple plan – to send a flying battery of artillery under Lieutenant Randolph Ridgely down the road at the enemy with infantry support on both sides. Ridgely became separated from infantry support and fell under Mexican musket and artillery fire.

General Taylor ordered Captain Charles Augustus May to charge the Mexican artillery with his squadron of dragoons. May found Ridgely's position and asked him where the guns were; Ridgely said he'd draw their fire.[18] After an exchange of artillery, May exhorted his men: 'Remember your regiment and follow your officers!'

At Resaca de la Palma, Grant described the heavy cannonade from the Mexican artillery: 'The [cannon] balls commenced to whistle very thick over-head, cutting the limbs of the chaparral right and left. We could not see the enemy, so I ordered my men to lie down, an order that did not have to be enforced.'[19]

Captain May led his squadron in columns of four and saber charged the Mexican artillery; during this charge, Captain May's bugler captured Mexican General Diaz de la Vega. As soon as the guns were silenced, Taylor ordered the infantry forward while Ridgely's artillery poured deadly fire into the ranks of the Mexican infantry. Arista's line collapsed and the Mexican retreat turned into a rout.

Charles Augustus May, a son of Maryland, is remembered to this day in the fourth stanza of the official state song 'Maryland My Maryland':

> With Watson's Blood at Monterey
> With fearless Lowe and dashing May.

The sabers used with great effect by Captain May's dragoons were the US Dragoon Saber 1840, seen in Figures 31, 32, and 35 above and below.

Chapter 6

The Model 1860 Cavalry Saber and the American Civil War

Although the Model 1840 was still in use at the outbreak of the Civil War, improvements in the quality of steel and demands by army officers initiated an update of the old wrist breaker, the slightly lighter and more ergonomic Model 1860 Light Cavalry Saber. Captain George B. McClellan, who would later command all Union forces early in the Civil War, made a report to the Secretary of War after an official tour of European arsenals, explaining the importance of the cavalry saber as a weapon while pointing out the French sabers of the period, although similar in appearance, were better balanced and lighter.[1]

Figure 34. One of the most common field pieces in use during the Civil War was the 12-pounder, often referred to as a 'Napoleon' after Napoleon III who was considered responsible for its adoption by the French for their army. It was widely copied in the United States.

Soon after the Model 1840 was initiated, the Ames Sword Company began making improvements in the quality of their sword steel. By 1845, homogeneous cast steel generally replaced shear steel – the process of repeated heating, rolling, and forging of blister steel.[2] The improvements allowed Ames and others to produce a new model that was almost an inch longer, but with a slightly lighter blade. In addition, the grip had a modest swell in the middle making it more ergonomic. (The swell in the middle of the grip is usually the easiest way to differentiate between the two sabers.)

Examples of what was officially called the new model but later known as the Model 1860 have been found dated 1858 and 1859; it is unknown why the new model was designated the Model 1860.[3] The Model 1860 was produced in large numbers, including a superficial update in 1906 – a steel guard replaced the brass guard (see Figure 47 below). It was the American cavalry trooper's sword for more than fifty years.[4] The Model 1860 Cavalry Saber was carried by Union and Confederate cavalry during the Civil War; then it was carried by US cavalry troopers in the West and in the Spanish-American War, including the war in the Philippines. Roughly 200,000 light cavalry sabers were produced for the Union during the Civil War, and continued in service into the early twentieth century.[5]

It should be remembered that both Union and especially Confederate officers might carry non-regulation sabers. Among Confederate soldiers the swords could range from high-quality European swords to crude, but serviceable, swords made by southern blacksmiths.

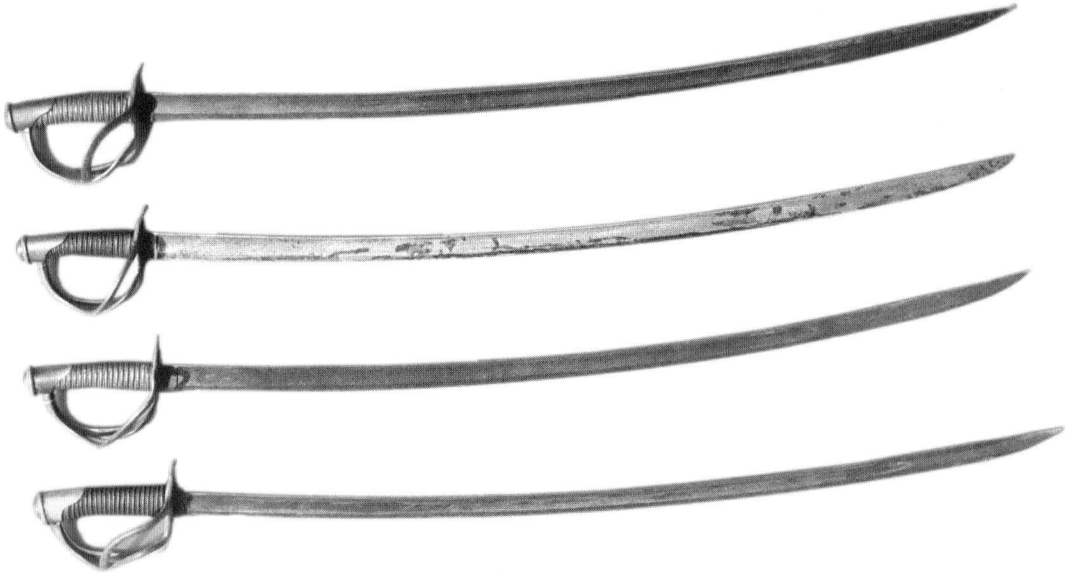

Figure 35. Four US Model 1840 Dragoon Sabers, from top to bottom: Tiffany manufacture. (*Fogle Collection*); German, C&J. (*Fogle Collection*); US manufacture with walnut grips indicative of Confederate use. (*Graham Collection*); and French manufacture with rooster mark on the ricasso. (*Graham Collection*)

At one end of the spectrum was Colonel Fred G. Skinner of the First Virginia who was described by Confederate General James Longstreet's aid, G. Moxley Sorrel, as:

> A fine swordsman and had brought from France a long, straight well-balanced double-edged cuirassier's saber. In his cups, the fine old Colonel would swear he should die happy could he have one chance to use that steel on the enemy. The chance came and Skinner was ready for it. At the second battle of Manassas, a battery of six guns was mauling some of our infantry horribly His regiment, the gallant First Virginia, was thrown at it, 'Old Fred' as the men affectionately called him, leading well in advance. Out flashed the French saber and he was among the gunners in a trice. His execution was wonderful; sabering right and left he seemed invulnerable, but down he came at last, just as his men swept over the guns in a fine charge.[6]

For examples of French cuirassiers' swords, see Figure 8 above.

Cavalry Swords in the Civil War

In America, the best parallel to the life of the Napoleonic Hussar could be found during the Civil War, particularly in the Confederate cavalry of the Army of Northern Virginia as led by J.E.B. Stuart. On two occasions: first from June 12–15, 1862, and again from October 9–12, 1862, Stuart led his entire cavalry command, more than 1,000 men, around the opposing Union forces commanded by General McClellan. Before the October raid, it was important for Stuart to have all the intelligence available. To that end, on June 6, 1862, Stuart, accompanied solely by his aide-de-camp, Heros von Borcke, went out at night to meet a spy. As they rode along in the dark, through enemy territory, von Borcke, suddenly checked the loads in his revolver and Stuart cautioned him that 'we would not employ firearms except in the last resort, and that in case of an encounter we must make use of our sabers.'[7] The saber carried by von Borcke would be mentioned throughout his memoirs, a long Damascus blade that he brought from his Prussian homeland.

Moxley Sorrel had occasion to meet von Borcke and described the ready-for-anything Prussian as follows:

> The German Von Borcke, who attached later to Stuart's cavalry made some reputation. He had just arrived and could not speak a word of English; was splendidly mounted on a powerful sorrel and rode well. He was an

> ambulating arsenal. A double-barreled rifle was strapped across his back, a Winchester carbine hung by his hip, heavy revolvers were in his belt, right and left side; an enormous straight double-edged sharp-pointed cuirassier's saber hung together with sabertasche to his left thigh and a short 'couteau de chasse' finished up his right. Besides his English army saddle bore two large holsters, one for his field-glasses, the other for still another revolver, bigger and deadlier than the others. When I next saw him he had discarded – taught by experience – all his arsenal except his good saber and a couple of handy revolvers.[8]

On the first ride around the Union Army, known as the Chickahominy Raid, Stuart led 1,200 cavalry and a section of artillery north out from the outskirts of Richmond. During the following three days, Stuart's reconnaissance in force would ride all the way around McClellan's army, cutting lines of communication and capturing valuable supplies and prisoners. Later that October, leading 1,800 cavalry, Stuart began west of Harper's Ferry and led his command north across the Potomac River into Pennsylvania to as far north as Chambersburg, then west and south into Maryland, finally crossing the Potomac again at White's Ferry, about 30 miles east of where he had originally crossed into Federal territory, and then to Leesburg, Virginia.[9]

At the skirmish at Brandy Station near Culpeper, Virginia, in 1862, not to be confused with the larger cavalry battle at the same place in 1863, von Borcke was involved in a cavalry action reminiscent of earlier wars:

> The enemy, as usual, received us with a rattling volley, which emptied several saddles; but a few seconds more and we were in the midst of them, and their beautiful lines, which we had so much admired, had broken into flight. I had the satisfaction here of saving my life by a magnificent blow upon one of my antagonists, who, at the very moment of firing at me received my full right-cut on the lower part of the neck, severing his head nearly from his body.[10]

Stuart and other Confederate leaders used cavalry to great effect, understanding that its role was primarily reconnaissance and covering the army's movements.

Brandy Station, June 1863

Perhaps the greatest cavalry battle of the Civil War was in early June 1863, near Culpepper, Virginia, at Brandy Station, ground that the Confederate cavalry had fought over before, but in 1863, it would be on a greater scale than ever before seen on this continent.

Soon after the Confederate victory at Chancellorsville, Confederate General Robert E. Lee began a northward movement of the army, with the cavalry screening this movement. On June 8, 1863, J.E.B. Stuart held a grand cavalry review for his commander and chief, which was interrupted by reports of Union Army activity on the north side of the Rappahannock River. It was fortunate that Stuart had concentrated all of his cavalry for the grand review, as it would turn out that he would need all of them.[11]

On the morning of June 9, 1863, von Borcke recalled being awakened at dawn by the sound of cannon fire and musketry, the Battle of Brandy Station had begun.[12] The 8th New York Cavalry under Colonel B.F. Davis crossed the Rappahannock River at Beverly's Ford and gradually pushed back the Confederate pickets. During the morning there were charges and countercharges by which the Union Cavalry began to take possession of the location of the Confederate position of strength on Fleetwood Hill. At the same time there was another threat to the Confederate position from Union Cavalry who had crossed the Rappahannock farther down at Kelly's Ford.

The heaviest fighting centered around Fleetwood Hill, a long ridge running roughly north to south, where hour after hour both army's cavalries charged and counter charged to claim the high ground.[13]

The turning point of the battle was the successful charge by Confederate cavalry led by Colonel P.M.B. Young of the Cobb Legion of Georgia. As Major Henry B. McClellan, who was present at the battle, recounted in his book, *I Rode With JEB Stuart*, 'Colonel Young led this charge straight up the Fleetwood Hill from the northeast. His regiment used the sabre alone. It was the turning point of the day and with the assistance of the 1st South Carolina Cavalry swept the Union soldiers from the heights.'[14]

By the following day the Union forces had retired to the north of the Rappahannock after the largest cavalry engagement in America up to that point,

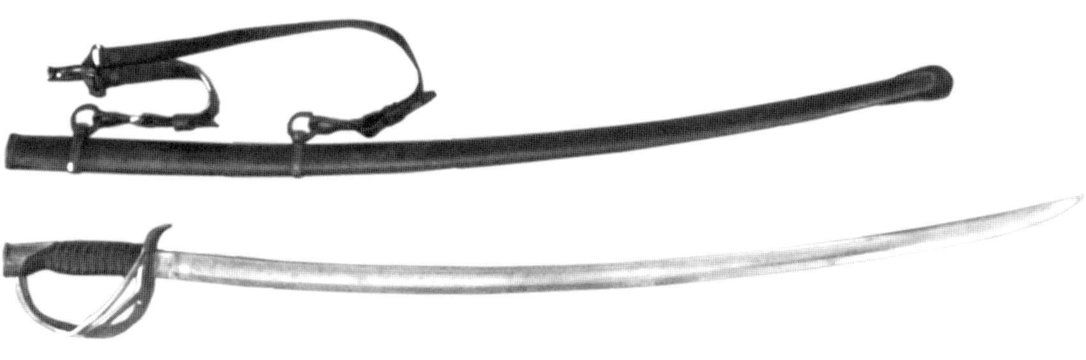

Figure 36. Model 1860 US Cavalry Saber and Scabbard, manufactured by the Ames Sword Company. (*Graham Collection*)

where 12,000 Confederate cavalrymen had contended with 15,000 soldiers of the Union Army. Confederate losses were just under 500, Union losses over 900, leaving Robert E. Lee free to continue his campaign into Pennsylvania.[15]

The vast majority of sabers used in the various charges during the Battle of Brandy Station would have resembled the example seen in Figure 36 above, which also provides a good example of the kind of scabbard and attachments used at the time.

Union General Judson Kilpatrick's Raid, Riding Around the Confederate Army in Atlanta

It would take the Union Army about three years to catch up with the Confederacy in terms of the proficiency of its cavalry, but eventually, Union General Phil Sheridan would ride around the Confederate Army at Richmond and Union General Judson Kilpatrick would ride around the Confederate Army in Atlanta with much the same effect as previous Confederate feats of arms, and Kilpatrick's raid is notable for its use of the saber.

In July 1864, Confederate General Joe Wheeler at Resaca, Georgia, cut off General Sherman's line of communication from the north.* Sherman responded by directing a similar raid to cut Wheeler off from his lines to Atlanta. Kilpatrick began his raid on August 18, tearing up the Macon railroad as far south as Jonesboro and was back in his former position, having ridden around Atlanta, by August 22. Sherman commenced his last movement on August 25 and Atlanta had fallen to the Union Army by September 2.[16]

Captain W.L. Curry, a survivor of Kilpatrick's raid, described the Union saber charge that occurred near Lovejoy, Georgia, when they were surrounded, facing five brigades of Confederate infantry and almost out of ammunition; Kilpatrick's brigades charged the enemy barricades in columns of four:

> [General Kilpatrick] ordered 'Draw Saber' and 4,500 sabers ring out as they are drawn from their scabbards, the reins are tightened, the horses are excited with nostrils extended as though they sniffed the battle far off … The command 'Forward!' is given the bugles ring out 'Trot!' 'Gallop!' 'Charge!' in quick succession and the columns swept forward under the spur with a yell, scaling fences, jumping ditches in that wild and reckless charge … [The] brigades urged their horses over the barricades, cutting right and left. Many [Confederate] prisoners had saber cuts on their hands, arms, and heads, and it is estimated that from six to eight hundred prisoners were sabered.[17]

* Wheeler would play an important role thirty-four years later, rejoining the US Army as a General Officer in Cuba during the Spanish-American War.

Figure 37. Model 1860 Cavalry Sabers, from top to bottom: Ames Sword Company, marked 1859; Ames prototype, marked 1860. (*Graham Collection*); Officers' Saber, made in Europe: Officers' Saber, marked Horstman. (*Fogle Collection*)

Figure 38. Four Union generals with examples of swords carried in the field during the American Civil War: Winfield Scott Hancock (seated), with division commanders, Francis Channing Barlow (left), David Bell Birney, and John Gibbon. Note that of these generals, Barlow alone is armed with the Model 1860 Cavalry Saber, the others carry the Model 1860 Staff and Field Officers' Sword, which was a symbol of rank but of little use as a weapon; the army would wait until 1902 to replace that frail weapon with a saber for all officers, for examples of the 1860 Staff and Field Officers' Sword and the 1902 Saber for All Officers see Figures 41 and 42 below. (*Courtesy of the Library of Congress Prints and Photographs Division LOT 4187, no. 167* (*P&P control number 2013647863, Washington DC 20540*))

The charge was a success, and the Union cavalry broke out to the east, towards Jonesboro, eventually regaining its place on the right of the Union forces north of Atlanta. By September 2, Atlanta would be in the hands of the Union Army.

The cuts delivered in the charge near Lovejoy were made by the New Model Saber, now known as the Model 1860, as seen in Figures 31 and 32 above.

After the Civil War and the publication of reports showing the small percentage of wounds and deaths caused by edged weapons, compared with those from rifled muskets and artillery, some officials questioned the need for arming cavalry with a saber. In 1878, General S.V. Bennet, the US Army Chief of Ordnance, questioned the need for swords and bayonets.[18] However, the Secretary of War did not act on the recommendation, and the sword and the bayonet were retained in the arsenal of American weapons.

A Brief Note on the Manufacturing of Swords

When we imagine the swordsmith, he might look something like the illustrations from Germanic myths, Siegfried with forge and anvil hammering out the blade with which he would defeat a dragon, or perhaps the Japanese swordsmith laboring to make a blade fit for his feudal overlord.

During the Middle Ages in Europe, while some swords were entirely the work of an individual craftsman, usually, the sword, with its scabbard and belt, was the work of at least five people: the 'bladesmith, the hilt maker, the sheather, the girdler and finally the furbisher, who assembled all the parts and usually sold the complete weapon.'[19] This cooperative effort in producing swords was prevalent in the numerous blade-making centers of medieval Europe, predating the division of labor postulated by Adam Smith centuries later.

In the Middle Ages the primary blade-manufacturing centers in Europe were: Pavia, Poitou, Bordeaux, Valencia, Passau, and Solingen.[20] By the Renaissance, blade production was primarily centralized in Toledo, Milan, and most importantly Solingen, in the Ruhr Valley of what is now Germany, and from where the Germans exported blades all over the world. The English would establish blade centers in Sheffield and Birmingham, which would produce the swords for British troopers, however, British officers would often purchase swords assembled in England but frequently with blades manufactured by the bladesmiths of Solingen.

The English settlers in colonial America brought their swords with them from England, but by that time much of the production of sword blades was centralized in Spain and Germany. Industrial centers known for sword making had factories with rolling mills that took red-hot billets of steel and ran them through rollers (see Figure 40 below) to extrude blade-shaped billets, which were finished into shape

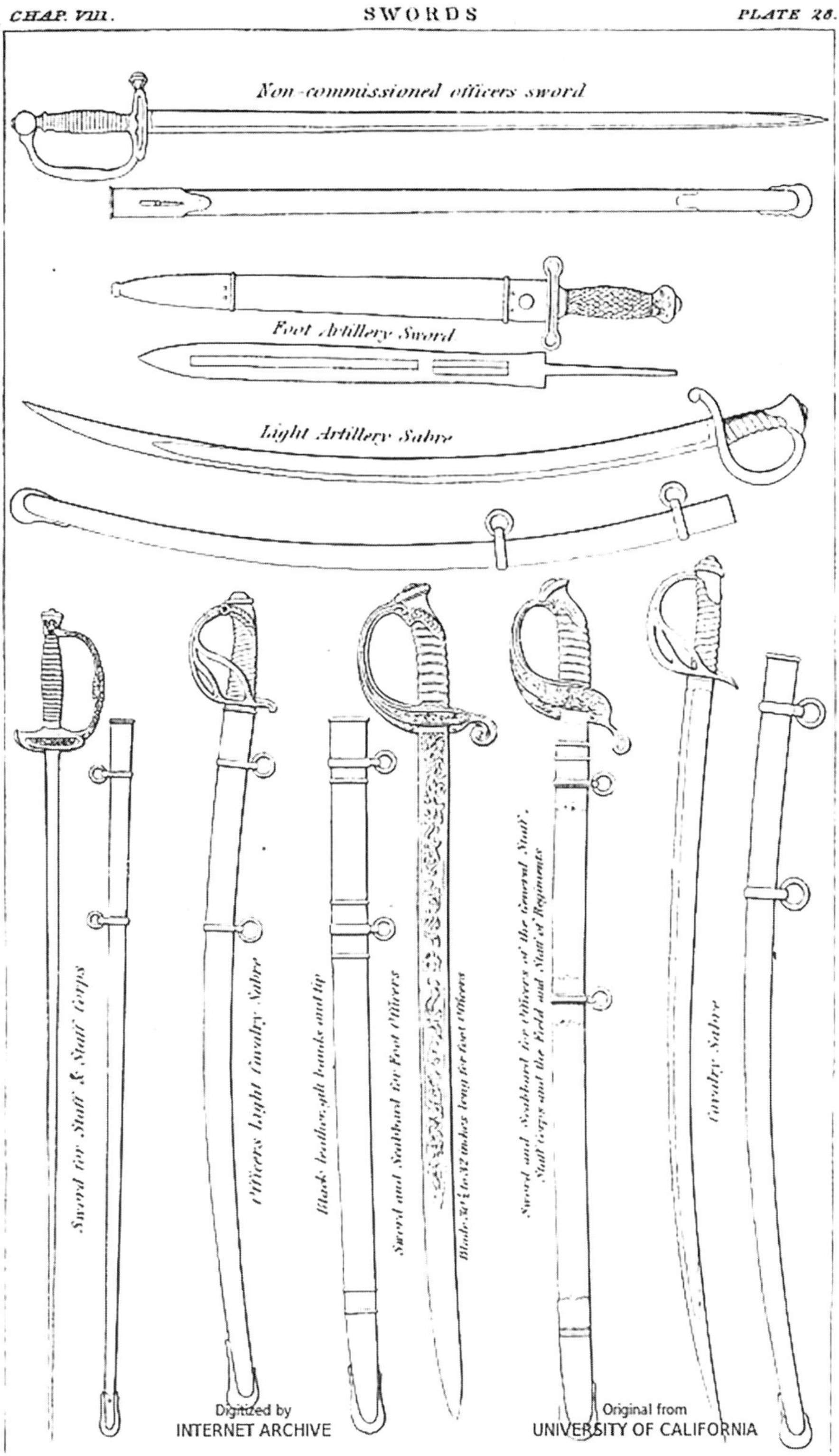

Figure 39. From the US *Ordnance Manual*, 1860. The Model 1860 Light Cavalry Saber is bottom right.

Figure 40. A rolling forge. (*Courtesy of Mr Andre Willms, W.K.C. Stahl-und Metallwarenfabrik*)

with trip hammers. The red-hot blanks were tempered, hardened, and ground into the final shape. Finally, craftsmen would polish and sharpen the blades.[21] Rolling out the blades allowed for greater uniformity in the various models of sword issued by the nations of Europe and eventually in the United States. The processes of production that arose during the Industrial Revolution for household goods were also applied to armaments, and the mass production of the sword-making centers became more efficient. At the Ames Sword Company, by the mid-nineteenth century, a team of workmen could cast sword hilts in bright shining brass and finish them quicker in one day than the same number of craftsmen could hammer them out over forms in a month. Although the rolling of steel appeared in Europe during the seventeenth century, by the nineteenth century, rollers had replaced hammers to such a degree that by the 1860s there were patents for manufacturing sword blades with mechanical rollers to protect inventors of these new improvements.[22]

Newsreel footage from the mid-twentieth century demonstrates how swords were milled during the Industrial Revolution – a particularly good example is from 1965, *Sword Making*, by Pathe.[23] The industrial techniques adopted by sword manufacturers enabled them to produce the vast quantities required by the armies of the American Civil War and later modern conflicts.

Chapter 7

American Empire

Transformation defined the United States after the Civil War, Atlantic and Pacific were soon joined by rail; industry and invention moved the United States from an agrarian to industrial power. However, the United States military, which was on par with the greatest European powers in 1865, saw a decline through the 1870s and 1880s that placed it behind some South American countries. American imperialism, born of the need to protect American interests and trade, would begin the revitalization of the US Army and Navy, and the resulting military actions would require improvements in US military equipment. During all of the American wars of the late nineteenth and early twentieth centuries, the saber would be part of the uniform of American officers and a trusted weapon of the US cavalry.

Perhaps five Americans could have claimed the greatest responsibility for the imperial expansion of US power – John Hay, Henry Cabot Lodge, Captain Alfred T. Mahan, Theodore Roosevelt, and Elihu Root.[1] Jointly and severally, they led the United States to the world stage. Perhaps, however, of this group of exceptional individuals, Mahan provided the most concise and popular argument for the strategic necessity for American imperialism.

The germ of Mahan's idea for a book on how the control of the oceans affected world history began as a result of his assignment to the South Pacific. During his frequent calls to Callao, Mahan would take a carriage to Lima and spend his time in the library of the English Club. When he read Theodore Mommsen's monumental *The History of Rome* he was struck by the question, what if Hannibal's Carthage had had control of the sea instead of Scipio's Rome? Wouldn't Hannibal have been spared the treacherous and wasteful journey over the Alps if he had control of the sea? Mahan returned to his duties at the Naval War College in 1885 (the year that George S. Patton was born) and gave a series of lectures on the effect of seapower to the Naval War College in 1886 – Theodore Roosevelt was one of his guest lecturers.[2] Pulling the lectures and his earlier ideas together, he wrote *The Influence of Sea Power upon History, 1660–1783*, which was published to acclaim and popularity in 1890. Mahan's sentinel work's popularity can be judged by the multitude of languages it was translated into and by those who advocated reading it, including Kaiser Wilhelm II. It was required reading in both the German and Japanese navies.

In this book, Mahan predicted the future construction of the Panama Canal and the incumbent strategic need to protect the sea lanes on both the Caribbean and Pacific entrances to the canal. In this he proved to be prescient, the United States would secure by force of arms the necessary strategic possessions – Cuba, Puerto Rico, the US Virgin Islands, Guam, and the Philippines within a decade of the publication of his book.

Of course, the understanding of the strategic necessities was important, but absolutely essential was the modernization and expansion of the long-languishing US Navy. Several scares about the power of certain South American warships sensationalized in west-coast newspapers, coupled with additional appropriations resulting from the wealth of new states in the west, helped to secure the funds needed to build all-steel warships for a modern steel navy. The destruction of one of these, the USS *Maine* in Havana Harbor on February 15, 1898, was the catalyst to the start of the Spanish-American War, culminating in the defeat of Spanish naval and land forces. By December 10 of that same year, under the Treaty of Paris, Spain ceded to the United States, the Philippines, Guam, and Puerto Rico, while renouncing sovereignty to Cuba – the security of the future of the Panama Canal was assured to the United States.

Mahan deserves some credit for the United States' success; since 1894, the Naval War College had drafted contingencies for a war with Spain, including naval action in the Philippines. In 1896 naval Lieutenant Warren Kimball's contingency plan for war with Spain specifically included the US Asiatic Fleet based in Hong Kong in action against the Spanish Philippine Islands.[3] However, equally important were the many actions taken by then Assistant Secretary of the Navy Theodore Roosevelt. Roosevelt benefited from his many connections, some made through the Metropolitan Club, just a few blocks up 17th Street from his offices at the State War and Navy Building. One of his most important connections there was Colonel Leonard Wood, whom Roosevelt would serve under in Cuba. The other important connection was a naval officer – George Dewey. During Secretary of the Navy Long's 1897 summer holiday, Roosevelt saw to Dewey's posting as the commander of the Asiatic Fleet. At his first opportunity after the destruction of the *Maine*, Roosevelt, on February 28, 1898, ordered the US Asiatic Fleet under his friend Dewey to 'SEE THAT, IN THE EVENT OF WAR, THE SPANISH SQUADRON DOES NOT LEAVE THE ASIATIC COAST.'[4]

The war in Cuba provided little opportunity for mounted cavalry action, partially because the army command limited the number of horses transported to those for senior officers, and of those, many died in transport. Consequently, the charge up San Juan Hill was made by unmounted cavalry, firing their Krag carbines at Spanish defenders. Even Lieutenant Colonel Theodore Roosevelt had to dismount when he came to a barbed-wire fence and continue on foot.[5] Decisive land victories, such as the one at San Juan Hill, and decisive naval victories at Santiago Bay and Manila Bay cemented the United States' victory over Spain. However, the victory at Manila Bay left the United States with a problem – whether just to occupy and control the port as a naval coaling and supply station or to take control of all of the islands; the United States opted for the latter course of action and in the Treaty of Paris Spain ceded the Philippines to the United States but many of the Filipino natives continued their fight for independence by switching allegiance from Spain to the United States.

The United States quickly seized the city of Manila from the Spanish defenders on August 13, 1898 in an assault worthy of comic opera. The commander of the Spanish forces, General Jaudenes, seeing his position as untenable contacted Admiral Dewey through the Belgian consul, and an agreement was reached to save the Spaniard's honor; when the Americans attacked from their position on the left of the Filipino line, opposite two opposing Spanish-held blockhouses, the Spanish resistance would end as soon after the American firing began as honor would permit.[6] The US forces quickly took possession of Manila.

However, the American soldiers denied the Filipino freedom fighters entry into the city, and thus began the friction between American soldiers and Filipino freedom fighters. Less than a month after the Treaty of Paris ceded the Philippines from Spain to the United States, on February 4, 1899, a patrol of soldiers from the First Nebraska Volunteers ordered some Filipinos to halt. They ignored the order and the Americans fired on them, thus beginning the Philippine-American War, which would continue unofficially until 1906.[7]

A major milestone of the conflict occurred in March 1901 when Colonel Frederic Funston envisioned a bold plan to capture the rebel commander Emilio Aguinaldo. Whether Funston had read Mahan or not, he realized that the US Navy had command of the seas and not the Filipinos. With eighty-one Macabebe scouts (Filipinos loyal to the United States), a Spanish soldier of fortune, Captain Lasaro Segovia, and two US cavalry officers, Captain R.T. Hazzard and Lieutenant O.P.M. Hazzard, who would be the colonel in command of US cavalry in France during the First World War, he sailed in the USS *Vicksburg* from Manila, south through the islands of Luzon's southern coast, into the Pacific and up the coast to the Casiguran Sound.[8]

After forced marches, where Segovia talked their way past insurgent outposts, they reached Aguinaldo's headquarters and fooled the guards that they were reinforcements, capturing Aguinaldo after a brief firefight, and ending a major phase of the war in the Philippines.

American Cavalry Saber Charge in China

Without the American occupation of the Philippines and military action against the Philippines insurrection, the US Army would not have been able to join the European and Japanese powers coming to the aid of the members of the foreign legations trapped in Beijing during the Boxer Rebellion.

As a result of the weakness of the Ch'ing Dynasty, a revolutionary group arose in China dedicated to ridding the country of Western influence, particularly Christianity. Calling themselves the Society of Righteous and Harmonious Fists, they earned from Westerners the nickname 'Boxers' because of their physical displays of martial arts. They began their murderous campaign on December 31, 1899, by beheading an English Christian missionary.[9] By June, the Boxers were gaining converts in the cities of Tientsin and Peking, while at the same time European and US warships gathered off the coast. A detachment of fifty US marines was sent to protect the American legation, which along with the representatives of other nations were besieged in the foreign compound in Peking. In China, the foreign embassies were known as legations, and they occupied a quarter in Peking south of the Imperial City and north and west of the Tartar city walls. The quarter included the diplomatic representatives of Austria, Belgium, Britain, France, Germany, Holland, Italy, Japan, Russia, Spain, and the United States, eleven nations in all.[10]

By July, the allied forces, including US marines and infantry, took the walled city of Tientsin, and on August 4, a contingent of 20,000 soldiers marched out of Tientsin to the relief of the Western legations in Peking. Among the US Army units sent to China from the Philippines were cavalry, who due to logistical challenges were in Tientsin without their horses. Although advocating taking part in the relief of Peking as unmounted cavalry, they were required to wait in Tientsin for their horses, which would need their care when they arrived.[11] Also, Tientsin was a strategic point up the Peiho River from the port cities of Taku and Tongku.

Although Cuba, and to a lesser degree the Philippines, had proven not particularly suitable for cavalry operations, cavalry was able to make contributions to the allied effort in China.

Becoming aware of a threat to the security of their hold on Tientsin, a British, Japanese, and American force led by British General Dorward left the city to meet the threat. The US cavalry component under Colonel Theodore J. Wint comprised 390 officers and troopers. In the early morning of August 19, 1900, the Bengal Lancers and the first squadron of the US cavalry under Captain A.P. Blocksom led an advance party; General Dorward followed with the British infantry, followed by the third squadron of the Sixth US Cavalry under Captain Forsyth. After covering about 4 miles in a south-westerly direction from Tientsin, the allied forces encountered outposts of Chinese nationalists and began to draw fire from them.

The allied cavalry dismounted and began to return fire. The allied line pushed back the Chinese outpost soldiers upon the Chinese main line whose right rested on the village of Pei-Tsang and whose left rested on broken difficult terrain, while the intervening countryside of about 1,200yd between the allied line and the Chinese was relatively flat open country. Only about 20 percent of the Chinese infantry had firearms, which Forsyth described as used energetically with very poor marksmanship; the remainder:

> Waived lances, spears and swords … going through evolutions that looked something like the stately measures of the minuet, except now and then they would jump up and down and brandish their weapons. Here and there one of these terrifying figures would collapse in a little heap on the ground and take no further part in the mystic pantomime; pierced by a Krag-Jorgensen bullet.

Forsyth's squadron was dismounted on the right of the line, the horses of the first squadron and the lancers were on the firing line, while the horses of Forsyth's squadron were about 300yd to the rear. As the British infantry was enveloping the Chinese right, the line of dismounted cavalry of which Forsyth's command was a part held a position about 1,200yd away from the parallel Chinese line. The Chinese attempted several infantry charges against the allied line but never came closer than 700 or 800yd from the allied line because of the superior marksmanship of the allies.

Once the attack by the British infantry on the Chinese right became apparent, the Bengal Lancers and the US First Squadron mounted and charged 'using first the revolver then the saber.'[12] Forsyth could see through his field glasses more than one Chinese infantryman struck down with the saber. This allied victory was one of the last actions in which the US cavalry would employ the saber. The sabers with which the US cavalry troopers struck down Chinese infantrymen were of the Model 1860 type, the same as those used during the American Civil War (see Figure 36 above). The fact that Forsyth uses the expression 'struck down' suggests that the saber was employed as a cutting weapon and not a thrusting weapon.

The victory over Spain and the acquisition of important foreign possessions – Puerto Rico, the Philippines, Guam, Hawaii, and Panama Canal Zone, and its construction, brought America to a new position in the world. The development of this place in the world was already underway with the US naval build-up in the 1880s. George S. Patton was born into an America that was very different from the nation into which his ancestors had been born, a nation ready to take its place as a world leader.

The republic's colonial expansion would have an effect on its ordnance, like Great Britain and France, whose soldiers served in far-flung outposts across the globe, and the nation's military leaders would have a keen interest regarding improvements in sword design. French and English officers serving in foreign posts found that brass sword guards were too soft and native blades might cut through them; they would often have their personal weapons made with steel guards – gilt when regulations required a brass finish. The leaders of the US military would continue to value the sword as an ancillary weapon and to support new swords to better meet the challenges of empire.

America Enters the Twentieth Century

In July 1902, the US Army adopted a new saber for all officers, a handsome saber based on French and Italian designs. It was felt that the Model 1902 would finally provide many officers with a useable sword, their only previous option being a sword that was purely a badge of office and in no real sense a weapon.

During the Civil War, Union infantry officers above the rank of captain who wanted a robust useable sword could carry the Model 1850 Staff and Field Saber rather than the 1860 Model, which had a delicate double-edged blade with an equally delicate guard (see Figure 41 below). However, the Model 1860 Staff and Field Sword became mandatory in 1872.

In February 1889, a board of infantry officers held the sword to be defective and unsuitable, too delicate in cross section to deliver a thrust and too light to deliver a cut. The board found the sword unsuitable for mounted work and requested permission to submit a sword that would meet the requirements of all infantry officers, however no action was taken on the report for thirteen years.[13] (At that time, cavalry officers carried a lighter version of the 1860 Troopers' Cavalry Saber, and mounted artillery officers had their own special saber.)

The adoption of the Model 1902 Saber for officers showed renewed interest in sword development, and its inspiration sprang from the Italian Cavalry Model 1888, a version of which is still carried by the Italian Carabinieri, today. However, the sword itself initially had its defects; its German silver guard and scabbard were

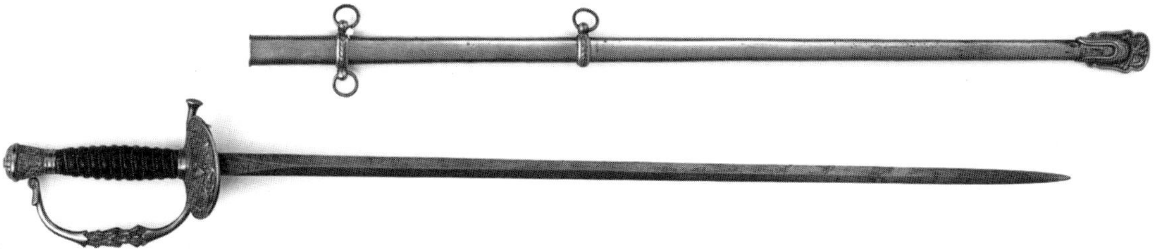

Figure 41. Model 1860 Staff and Field Officers' Sword. (*Courtesy of the Smithsonian Institution, Museum of American History*)

Figure 42. The Model 1902 Saber for all officers, the replacement for the Model 1860 Staff and Field Sword, SPAR 6594. (*Courtesy of the Springfield Armory Museum*)

found to be too soft, and in December 1902, regulations required the guard and scabbard to be made of steel in a dull nickel finish.

In 1904 the War Department reported that the supply of light cavalry sabers from the Civil War had practically been exhausted, and the department ordered 20,000 replacements. The campaigns in Cuba and the Philippines certainly took a toll on the stocks of Model 1860 Sabers.

However, the decision to continue arming officers and men with sabers was not automatic. Some officers proposed discarding the sword as part of normal field equipment, suggesting that cavalry troopers should be armed with three to four revolvers per man instead of arming them with revolvers and sabers.[14]

The motion to discontinue the saber was overruled, and an army special committee recommended that a new saber should be both a thrusting and cutting weapon and that different sabers were appropriate for mounted soldiers and infantry. The committee further felt that in the eventuality that an officer would have to defend himself with his saber, it probably would not be against another officer similarly armed but against an infantry soldier armed with a rifle and bayonet, with which a sword cannot usually compete.[15]

The consideration of officers protecting themselves with their swords against infantrymen with rifles and bayonets was the result of reports from US Army observers during the Russo-Japanese War of 1905.[16] Consequently, it was decided that the new sword should be quick and not heavy on the point. They recommended a grip that expanded toward the base, so that 'it tightens in the hand as the saber goes forward in the cut, all sides of the grip increasing in dimensions toward the pommel.'[17] The result of the committee work was the experimental saber of 1906.

The following year, the War Department would recommend developing a new saber, and an experimental saber was designed, which would have a light handy blade and the grip 'increasing in dimensions toward the pommel.'[18] The description is readily apparent in examples of the 1905/1906 experimental sabers available in collections today (see Figure 43 below).

At this time, the only sword manufacturer in the United States was the Ames Sword Company, with whom the Springfield Armory had shared examples of

Figure 43. Model 1906 experimental saber and close-up of the hilt of a 1906 experimental saber, Officers' Model, note the 'S A' and the flaming bomb above the '1906' on the ricasso. (*Courtesy of Tom Wiltzius and John Siekierka, Springfield collector's guide*)

the experimental saber. In total the Ames Sword Company produced 1,039 of these sabers, all of the experimental sabers that were produced for troopers were manufactured at the Ames factory. However, the Springfield Armory manufactured some of the 1906 experimental sabers for officers; Figure 43 above shows one of these rare items.

The 1907 army regulations for saber exercises contains illustrations for saber drills on foot and mounted; the saber used in the illustrations is the 1906 experimental saber. The on-foot exercises are similar to the sport fencing instruction for sabers at the time, with notes on advances, retreats, parries, and attacks.[19]

The 1906 Troopers' Saber

The Ames Sword Company provided the army with another 18,961 sabers based on the Model 1860 Cavalry Saber from the Civil War, updated by replacing the brass guard with a more durable drop-forged steel guard. The blade remained virtually the same as the Model 1860 except that it was manufactured in Germany to enable the company to fill the order for the manufacture of the experimental saber blades.[20] For an example of the 1906 update of the cavalry saber in use since the Civil War, see the sabers in Figures 44 and 45 below, all of which are stamped '1906,' however they were ordered into existence in 1905. The scabbard was modified from the 1860 version by moving the bands closer together for better balance while riding. This was the cavalry saber used by troopers when Patton became a cavalry officer.

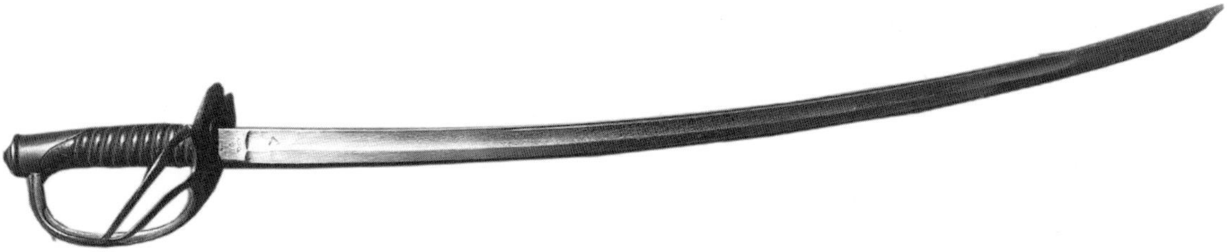

Figure 44. US Model 1906 Troopers' Saber, marked Ames, the blades were made in Solingen, Germany. (*Author's Collection*)

It may seem anachronistic that the US Army of 1913 would design, adopt, and produce a new cavalry saber in the same year in which the first US Army airplane squadron was formed. At this time the army was already moving troops and equipment with trucks, and soldiers attained the deadliest marksmanship, up to that time, with modern bolt-action rifles firing high-velocity smokeless ammunition. Cavalry troopers and officers carried the Model 1911, 45-caliber auto-loading sidearm, which the army would retain right up into the 1990s. Nonetheless, the army continued to pursue a goal of developing a better cavalry saber, and there were those in the army who felt that a better saber should be developed along the lines of the straight-bladed cavalry swords adopted by European armies.

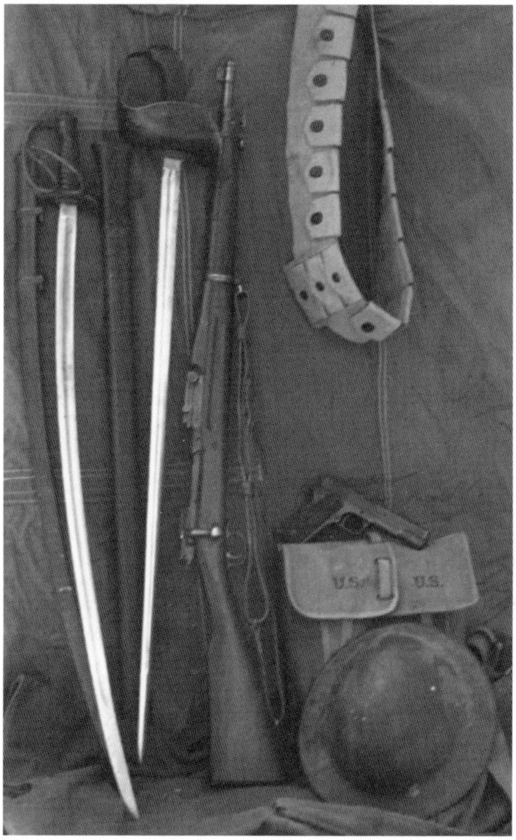

Figure 45. Early twentieth-century US cavalry weapons, from left to right: Cavalry Saber of 1906 and Model 1913 Cavalry Sabers and Scabbards. (*Author's Collection*) Model 1903 Rifle and Model 1911 Pistol. (*Fogle Collection*)

Other forces were at work that would help to provide the resources the US military would need for the challenges of the twentieth century. Until February 25, 1913, with the passage of the sixteenth amendment to the Constitution of the United States, the government was prevented from collecting income taxes directly from its citizens' personal income. However, with the passage of the amendment, the government could tax individual incomes, and the income tax became law on October 3, 1913. The Commissioner of Internal Revenue reported total revenues for 1914 of $380,008,893.96, the greatest amount in the history of the bureau, up 10 percent from the previous year's total. Included in the total was the new tax on individual income, raising $28.3 million, or about 7.4 percent of all revenue for the year.[21]

Leaders in the United States anticipated a greater role for the republic in world affairs and recognizing the related fiscal responsibilities passed revenue measures similar to the excise taxes on liquor and tobacco that had helped fund the war effort against Spain in 1898, in addition to the new taxes on incomes of individuals and corporations. With the First World War in its second year, the United States enacted the omnibus Revenue Act of 1916, which boosted income tax rates and added a new estate tax, and taxes on stock dividends and profits of munitions manufacturers.[22]

Figure 46. Two twentieth-century US cavalry sabers, top: the US Saber Model 1913 designed by George S. Patton, marked Springfield Armory, 1913; bottom: the 1906 Saber (basically a Model 1860 updated with a steel hand guard), marked Ames Sword Company, 1906. Interestingly, all of the blades of Model 1906 Sabers, although marked 'Ames' on the ricasso, were made in Solingen, Germany. (*Author's Collection*)

Another Experimental Model

In the 1912 War Department report the Chief of Ordnance announced that a number of experimental sabers were being manufactured to develop a suitable saber for officers and enlisted men of all arms of the service. 'The saber selected will be tried out in field tests.'[23] Those tests were conducted by two troops of the Eleventh Cavalry testing equipment supplied from the Rock Island Arsenal in Illinois.

In 1911, the Cavalry Equipment Board met at the Rock Island Arsenal to produce a report making recommendations for various updates, including a replacement

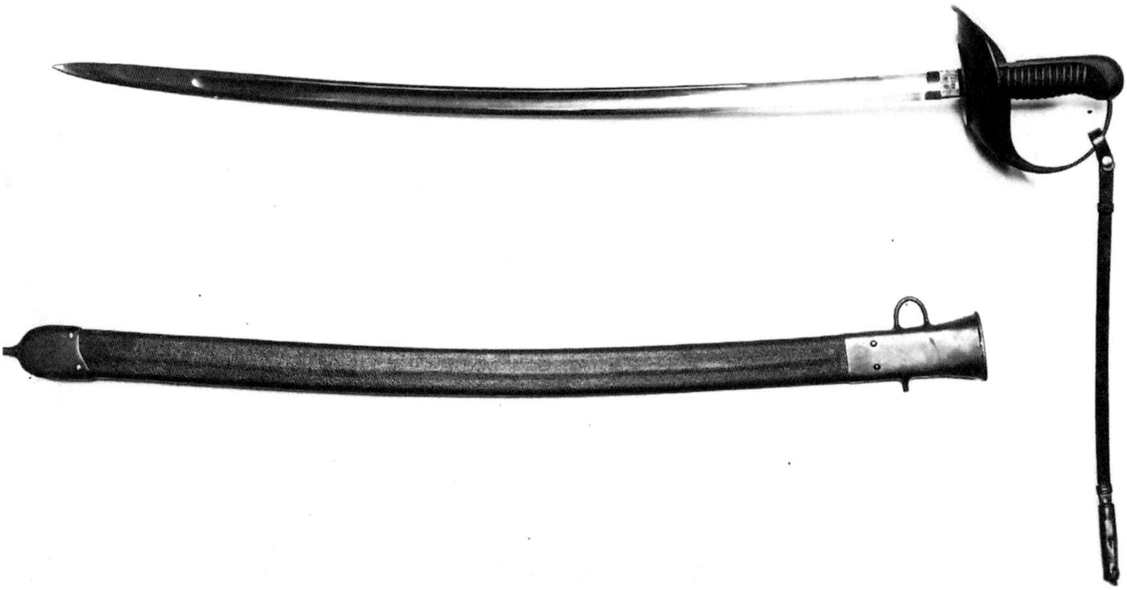

Figure 47. Experimental Model 1911 Cavalry Saber, manufactured at the Springfield Armory. (*Image from the Cavalry Equipment Board Report, 1912, courtesy of the United States National Archives*)

of the 1906 update of the US Model 1860 Saber. The board had discarded the 1906 experimental design and adopted a robust cut-and-thrust sword with a more ergonomic grip and a protective, bowl-shaped hand guard of sheet steel. Like European swords of the late nineteenth century, the sword has a metal backstrap on the handle and the guard provided excellent hand protection.

The 1911 experimental saber closely resembles the 1860 Model Spanish Light Cavalry Troopers' Saber (Sable para Tropa de Caballería Ligera, modelo 1860 'Prusiano'), a sturdy cut-and-thrust saber which was still in use during the Spanish-American War, making examples available to the US Army of that time. The Springfield Armory manufactured fewer than a couple hundred of the 1911 experimental sabers; today, they are extremely rare.

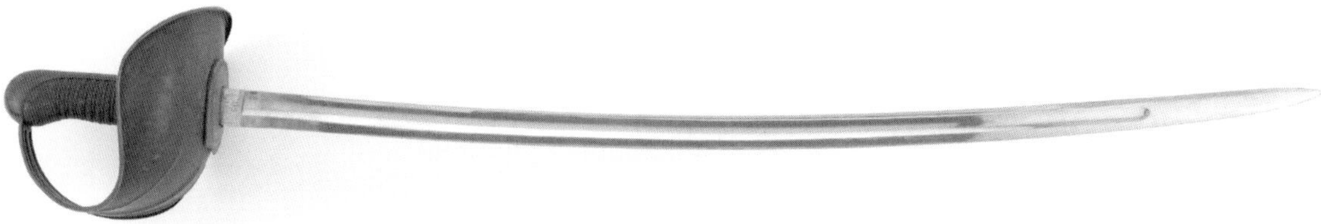

Figure 48. Experimental Model 1911 Cavalry Saber, note that this version has three strengthening grooves on the guard, an improvement that would be carried over to the Model 1913. (*Courtesy of the Springfield Armory Museum*)

The Army Cavalry Equipment Board issued a report in 1912 and described the 1911 experimental saber as favorably meeting what it felt would be the needs of cavalry in the field – a robust cut-and-thrust sword:

> The saber is the most effective weapon with which Cavalry in masses can meet Cavalry in masses, in mounted combat. The saber designed by the Board is a cut and thrust weapon, the blade being 31 7/10" long to which should be added the length of the grip 6 ¼". It weighs two lbs. In endeavoring to produce a saber well adapted to both cutting and thrusting, the Board combined the best qualities found in sabers especially designed for cutting and other sabers especially designed for thrusting. The combination attained is believed to be a very fortunate one. The point of the blade is on the median element thereof, thus favoring accuracy ... the point is double edged in order to facilitate penetration. The cross section of the blade has been designed so as to best combine strength and proper balance with an edge which can be easily kept sharp; the Board's recommendation being that this saber be issued sharp and kept in that condition. The steel guard is dark finished (anchoride), presents an unbroken surface to cuts and thrusts and gives ample protection. The grip is of wood shaped to the closed hand and covered with shark skin, wired down. This saber is a service weapon and it is recommended that paragraph 1544 A.R., 1910, be amended so that this saber and its scabbard can be drawn by officers from stores as is now the case with the rifle, revolver, etc. It is recommended that the present officer's saber [the Model 1902 previously described] and scabbard be retained for use in garrison.[24]

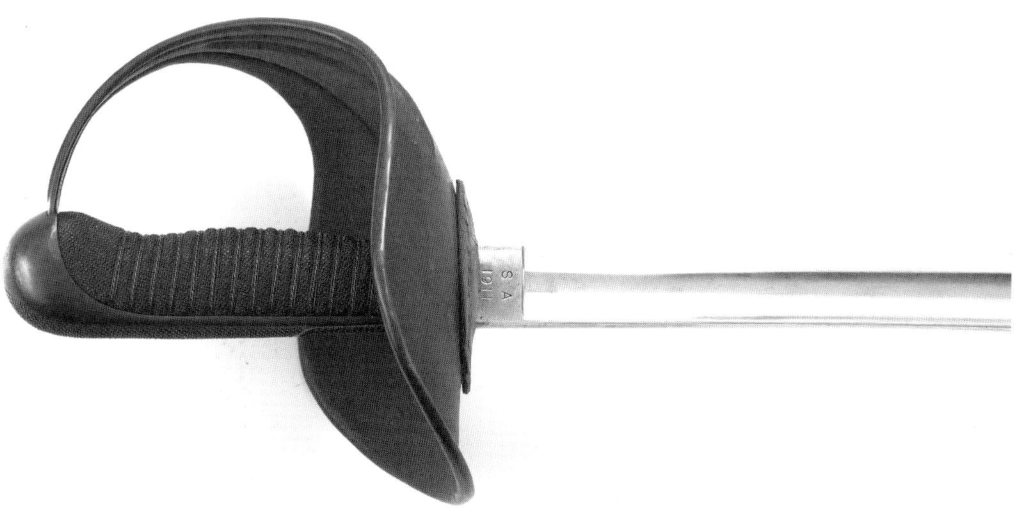

Figure 49. Close-up of the hilt of the 1911 experimental model, SPAR 6625. (*Courtesy of the Springfield Armory Museum*)

Figure 50. US Cavalry trooper, mounted and with full kit including the experimental Model 1911 Cavalry Saber, from the 1912 Rock Island Armory Cavalry Equipment Board Report. (*Courtesy of the United States National Archives*)

The 1911 experimental saber, serial number 120, had a checkered backstrap and thumb depression and a solid bowl guard with three stamped-in strengthening ribs.[25] These improvements would be carried forward eventually into the Model 1913 (see Figures 47, 48, and 49 above).

Beginning with the first mounted troops in the Continental Army, during the War of 1812, the Civil War and right up to 1913, the curved saber was the only type of sword for the US cavalry; the closest thing to a thrust-centric saber in the US arsenal would be the slightly curved 1833 Dragoon Saber, which with its pipe-back design was well suited to both thrust and cut (see Figure 26 above).

It is important to point out another significant small arms development that occurred in 1911, the US Army adopted the Colt semi-automatic pistol in 45 ACP; a pistol so modern that it would remain the US military sidearm until the early

Figure 51. Spanish Model 1860 Cavalry Saber, marked Toledo 1873. (*Courtesy of the Smithsonian Institution, National Museum of American History*)

1990s. Not surprisingly, in 1912, army regulations required dismounted officers in the field to be armed with a pistol instead of a saber.

The future of the sword in the army would be in the cavalry, and in 1913 the United States would adopt a modern straight sword for the cavalry because of one man, George S. Patton.

Chapter 8

Patton

George S. Patton, famous to the world as one of the great battlefield commanders of modern history, was descended from an illustrious line of American military heroes. His great, great, great-grandfather was Revolutionary War hero Hugh Mercer, a general who served under his friend George Washington; he was killed in action in Washington's campaign to capture British supplies at Princeton, New Jersey.[1] Patton was the third in line in his family to bear that name and the third to attend the Virginia Military Institute (VMI).

Figure 52. Patton demonstrates the proper use of his new cavalry sword in the charge, leaning well forward gives the greatest reach with the point while the rider's head and torso are protected by the horse's head and neck. Notice how the knuckle guard is pointed toward the sky; the sword hand had to be rotated because the grip was in line with the blade. This position may also have helped extricate the blade from a disabled opponent, and, in addition, it follows the French school of cavalry swordsmanship. (*Courtesy of the Library of Congress, Patton Papers, Box 60*)

His paternal grandfather, George Smith Patton, was class of 1852 at VMI and rose to the rank of colonel in the Confederate States Army. Although he had been recommended for promotion to brigadier general, the promotion did not come through before his death on the battlefield. He was struck by fragments of a Union shell as he stood in the stirrups of his horse on a street in Winchester during Jubal Early's defense of the Shenandoah Valley, dying of his wounds on September 25, 1864.[2]

The second George Patton also went to VMI and graduated in 1877, but had to forgo a military career to take care of his mother and other members of his family who had settled in California. He practiced law, was involved in land development, and participated in local politics, eventually running as the Democratic nominee in California's 1916 Senate race, which he lost.

George S. Patton was born in 1885, entered VMI, but left to finish his education at the United States Military Academy at West Point, graduating in 1909.

Patton's most diligent historian, Martin Blumenson, began his major work, *The Patton Papers*, by writing, 'Everything that everyone has ever said about George S. Patton, Jr. is probably true.'[3] Going on to describe him as a much more complex figure than 'the blustering egomaniac portrayed by the war correspondents,' Blumenson provides the reader with a full and honest account of Patton's life.[4]

Martin Blumenson, who died in 2005, was no ordinary historian; as an army officer in the European theater during the Second World War and later in the Korean War, he saw war first-hand. Afforded the great good fortune of being the first historian given full access to the private papers of the late General Patton, well before this archive was made available to researchers through the Library of Congress, he became the foremost author on Patton. Each volume of *The Patton Papers* weighs in at almost 1,000 pages, providing an honest assessment of one of America's most important generals. (The entire archive of Patton's papers in the Library of Congress comprises seventy-eight regular containers plus twenty-three oversize containers: the regular containers are 14in by 8½in by 4½in.)

Blumenson captures Patton's complex character: 'driven by ambition, tortured by self-doubt.'[5] The author shows an insecurity that began during his school days and continued after his five years at West Point. Only through tremendous effort and self-discipline was he able to graduate in the top half of his class. Later he would attain academic distinction at both the Command and General Staff College and the Army War College.

Although Patton was frequently inconsistent – he could show the prejudices of a descendant of Confederate leaders and yet bravely prevent the lynching of an African-American soldier accused of rape; the incident was at Fort Riley, Kansas, in 1916 when he was the only officer on duty. He informed the local citizens that if their intention was to lynch the soldier, it would be over Patton's dead body. The soldier was later proven innocent.[6]

With these inconsistencies in mind, there was one area where Patton was unswerving – his pursuit of perfection as an army officer. As a young boy he wrote a poem titled 'Forward Knight,' and announced to his family that he would become a soldier – a cavalry officer in the army being the closest thing to being a knight.[7] During his life he would eschew the goals of many young men, such as making a fortune in business, and focus on becoming the twentieth century's equivalent of the perfect knight. Patton would participate in the modern parallels of knightly tournaments – polo matches, horse races, fencing tournaments, and eventually the modern pentathlon in the V Olympic Games in Stockholm in 1912.

Patton had the good fortune to know and be on good terms with several Chiefs of Staff of the US Army. Early in his career, Patton was a junior officer on the staff of General Leonard Wood, army Chief of Staff, who, although a medical doctor with a degree from Harvard Medical School, was, during the Spanish-American War, the commander of the Rough Riders in Cuba – Theodore Roosevelt was the better known colonel in that regiment. Before becoming Chief of Staff, Wood served as the Governor General of the Philippines.

In his capacity as a staff officer, Patton had the opportunity to make his opinions known to the most senior officers in the army, including his boss, the Army Chief of Staff. In 1909, the year he graduated from the United States Military Academy at West Point, Patton married Beatrice Ayers, the daughter of a rich Boston businessman, her wealth and support helping him immeasurably in his army career. She was fluent in French and translated French cavalry training materials for her husband. In Washington DC her family's money helped Patton maintain a social life that in turn assisted his career. His father-in-law advocated that Patton follow

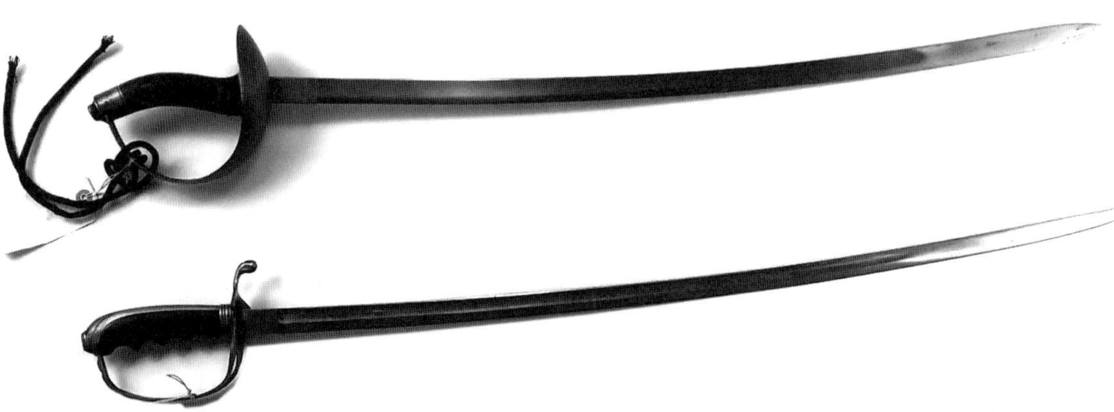

Figure 53. Sabers belonging to General Leonard Wood, top: Spanish Cavalry Saber; bottom: US Model 1902 Officers' Saber. (*Courtesy of the Smithsonian Institution, National Museum of American History*)

a career in business rather than the army, but Patton would not swerve from his devotion to the US Army.

Patton had distinguished himself in athletics at the academy, playing football and competing in track and field and fencing. It is important to note that a movement began shortly before Patton's birth and saw fruition during his early life – the shift in interest from dueling to modern sport fencing.[8] The sport fencing saber was developed in the 1870s and its rise in popularity was rapid, particularly among the officer class. Modern fencing technique became part of the saber manual of many nations including the United States. Although practice fencing and competition had existed for some time, the regulation and standardization of the modern sport fencing weapons was given impetus by the modern Olympics, beginning in 1896. The three weapons – foil, saber, and épée – were standardized so that, excluding electronic scoring, they resemble the weapons used today.[9] The weapon Patton competed with in Stockholm was the modern sport fencing épée, although he and others, up until the 1940s, still referred to it as the dueling sword.[10]

In 1912, Patton won a place representing the United States in the modern pentathlon at the Olympic Games held in Stockholm, Sweden. These games attracted more attention and spectators than any of the previous Olympic Games and brought fame to such great athletes as the decathlon champion Jim Thorpe, often considered to be the greatest athlete of the first half of the twentieth century.*

The modern pentathlon was a new event, the idea of Pierre, Baron de Coubertin, the founder of the modern Olympic Games. He believed there should be an event that would encompass the physical demands of the modern army officer – fencing, horsemanship, pistol marksmanship, running, and swimming; as Patton later wrote, 'five competitions intended to test the fitness of a perfect man at arms of the present day.'[11] George Patton was well suited to the event, having participated in ocean swimming events in his home in California. He was an avid horseman, fencer, and pistol shot, and he ran track at the academy.

* Interestingly, soon after his Olympic competition and success with developing the new cavalry sword, Patton received a letter informing him that he was unanimously elected as a member of the American Olympic Committee for the Olympic Games to be held in Berlin in 1916; the letter from Olympic Committee secretary James Sullivan was dated June 24, 1914 – four days before the assassination of Archduke Franz Ferdinand in Sarajevo. But on June 24, 1914, no one knew that most of the world would be involved in the most costly war the world had known, up to that point, and that the postponement of the Olympic Games would pale in to insignificance in the shadow of the horrors that would claim more than 10 million lives. And not being prescient, Patton following up on his Olympic experience, would focus on the development of a new sword for the cavalry. In a letter to his wife, shortly after the Olympics, he expresses hope that the army will adopt his design.

Patton did well in all of the events, including being the only competitor to beat the French fencer, Jean de Mas Latrie, considered the best swordsman competing in the modern pentathlon.[12] However, Patton finished in fifth place in the pentathlon, the first four places all going to members of the Swedish team. In fact, Patton was the only non-Swede in the top seven places.[13] The fact that he was the only competitor to place among the competitors from the host country, whose team had unmatched depth, reflected well on Patton and was recognized by his superiors in the army.[14]

Oddly, the reason Patton ended up in fifth place instead of receiving a medal was because of the pistol shooting event, which consisted of firing 20 rounds in groups of 5 shots at 4 targets. He scored well on all but the second set, where he scored 3 '10s' and 2 complete misses, his only complete misses of the competition. Right before the opening of the games, Patton scored 197 out of 200 in practice.[15] However, his poor showing in the pistol event during the games has caused some to theorize that the two 'misses' had passed through the hole made by his 3-shot grouping at the center of the target.[16]

Patton did not question the scoring. He wrote: 'There was not a single incident of a protest or any unsportsmanlike quibbling or fighting for points which I regret to say marred some of the other civilian competitions at the Olympic games. Each

Figure 54. Epée fencing as part of the modern pentathlon event at the V Olympic Games, Stockholm, 1912. Jean de Mas Latrie is seen on the left and George S. Patton on the right. (*Courtesy of the Library of Congress Manuscript Division, Patton Papers, Box 5*)

man did his best and accepted what fortune sent like a true soldier.' [17] It was as though Patton was competing with other knights of his station at a tournament during the highpoint of the Middle Ages.

The fencing in the modern pentathlon in 1912 was done with a robust épée, a thrust-only weapon where touches could be scored only with the point, unlike the saber which allowed cuts. Unlike the foil, a light 'weapon' only suited to sport, the fencing épée was like the swords used for dueling at the close of the nineteenth century. Additionally, unlike the foil, where only the torso was target, in épée, everything is target, from head to foot.

Patton was recognized in the official report by the Swedish Olympic Committee in its description of the four top fencers in the modern pentathlon: 'Among these fencers, Gronhagen, Sweden, and Patton, USA, distinguished themselves by calm skill and rapidity they showed in the use of their weapons as well as for the immediate advantage they took of the weak points of their opponents.'[18] Even before the Olympic Games, Patton was well placed to forward his opinion about a new thrusting sword for the cavalry, because he was on the staff of the chief of staff of the army and because he was a respected amateur fencer, but his recognition as a swordsman in the Olympics gave him a heightened platform from which to advance his ideas.

Chapter 9

Patton's Report on the Olympic Games and the Birth of the Model 1913

In the official report that Patton wrote to his army superiors about his part in the 1912 Olympics, he not only describes his participation, but uses his report to make an eloquent case for a new cavalry sword for the US Army. He forcefully opines that the ideal sword for the modern cavalry should be straight, not curved. 'The curved saber is a hybrid, being a cross between the rapier and the scimitar and having the good qualities of neither.'[1] Surely Patton was referring most specifically to the US cavalry saber in use at the time, a steel hilted version of the saber carried during the Civil War.

Patton approved of the épée, or as he described it:

> The duelling sword is a weapon with a 34 ½ inch blade of triangular cross section 2 inches in circumference at the hilt and tapering to a point. It has a bell guard five inches in diameter and weights from 1 ½ to 1 ¼ lbs. It is the rapier of history and the ancestor of all swords … It is a very excellent adjunct to teaching the use of the saber because it develops the use of the point wherein the true value of the saber rests.[2]

In his writings Patton equates the rapier with the finest of sword designs. Rapiers, usually long, double-edged with a deep fuller and designed primarily for thrusting, were clearly in Patton's mind when he designed his cavalry sword; the blade of the Cavalry Saber Model 1913 has much in common with many rapier blades.

In his report, Patton pointed out that he had recently competed in saber at the National Championships in the spring of that year in New York, but after working on his fencing at the French cavalry school under Adjutant Clery, he became a disciple of the French style repeating what he had learned: 'La pointe … toujours la pointe!'

Using the report to make his argument for a new type of American cavalry sword and sword manual, Patton enumerates why a straight thrusting cavalry sword is better:

> while the cutting saber is most effective in cutting about 9in from the point, with a straight thrusting sword the effect begins at the very point; the point is more difficult to parry; when reaching for a man with the point, you are partially covered with the guard and are in a position to parry; the point is far more deadly; the point will arrive before the cut …

Patton extolled the virtues of the French cavalry training and its master of the sword, Adjutant Clery:

> Clery's style of teaching is quite different from that used in America and in my opinion far better. It minimizes effort and is safer. I have fenced a great deal with the saber and in fact fenced in the National Championship in New York in the spring of 1912 that I believe my opinion is based on knowledge, not a desire for something new … The French use the point far more … [The] mounted position of Charge Saber is as follows: the trooper leans forward along the horse's neck, slightly to the left side of it so that only his eyes appear over the mane; his right hand grasps the saber with fingernails down and guard turned slightly to the right, point far to the front and above the level of the horse's ears … the arm is not drawn back on impact, there is no jabbing, the sword is held still and the horse does the work … there is one point parry to the right, there is no parry for a cut because a cut would never reach; the man turning to cut would have six inches of blade in him before he could start his stroke … the French recruit wants to use the edge just as much as ours do but it is drilled out of them and 'La point … toujours la point' is put in its place;[3]

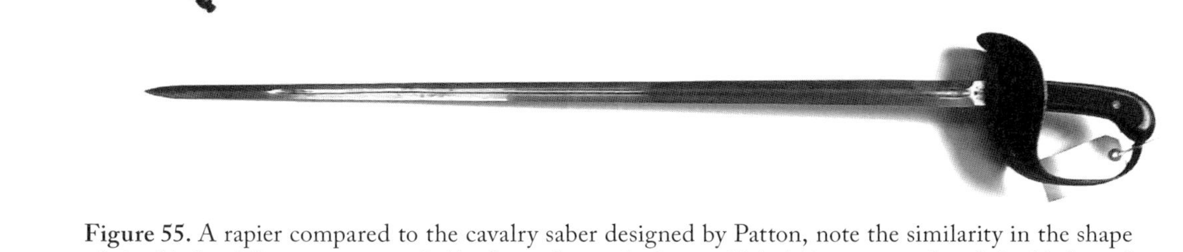

Figure 55. A rapier compared to the cavalry saber designed by Patton, note the similarity in the shape of the blades. The rapier is probably Spanish, marked in gold '1517' on one side of the blade and 'HORTA' on the other, blade length 43.5in, overall length 50⅝in. It was a gift from Ralph Packard to the Smithsonian in 1943. (*Courtesy of the Smithsonian Institution, National Museum of American History*)

These ideas would be repeated in the army *Saber Exercise, 1914* manual, which Patton authored.

Patton concluded his report by pointing out that the ancestor of the saber, the scimitar, was used in conjunction with a small shield and therefor it was never intended to be a weapon with which one could parry; further, the saber by having a less curved blade doesn't cut as well and is difficult to give point with, which he surmises is why 'the French, English and Swedes and … most other nations are adopting straight swords.'[4] He felt that it would be suicidal to engage enemy cavalry who were armed with modern straight swords, with US troopers armed with an update of the saber carried in the Civil War.

Patton understood that the sword, in the early twentieth century, was purely a weapon for cavalry. During the campaign on the Mexican border:

> He talked incessantly about his saber design, so much so that on the Punitive Expedition his men teased him for killing a rattlesnake outside his tent with a pistol instead of a sword. He saw no reason to laugh. Of course he'd used a pistol – he was on foot at the time and a saber was to be used strictly on horseback.[5]

In composing a response as an officer of the Chief of Staff, he wrote:

> [In] the regulations for the Uniform of the United States Army, published March 25, 1912, will be found a paragraph which requires dismounted officers to carry the pistol and not the sword when in the field. The idea of carrying a light rifle has been considered by the Department from time to time, but it is generally thought that were an officer armed with a rifle he would pay more attention to his personal combat than he would to looking after the interests of his men.[6]

Figure 56. French Army Adjutant Clery, right, who trained Patton in fencing both before and after the Olympic Games in 1912. (*Courtesy of the Library of Congress, Paton Papers, Box 74*)

Further, Patton made a convincing argument for focusing on the sword thrust: the point thrust at an opponent arrives before a cut, and a sword thrust is much more likely to kill, disable, or unhorse an opponent than a cut. 'In the charge, the point will always beat the edge, it gets there first. It is at least five times more deadly'[7]

And although Patton understood the value of cutting swords like the scimitar and the early sabers, he felt

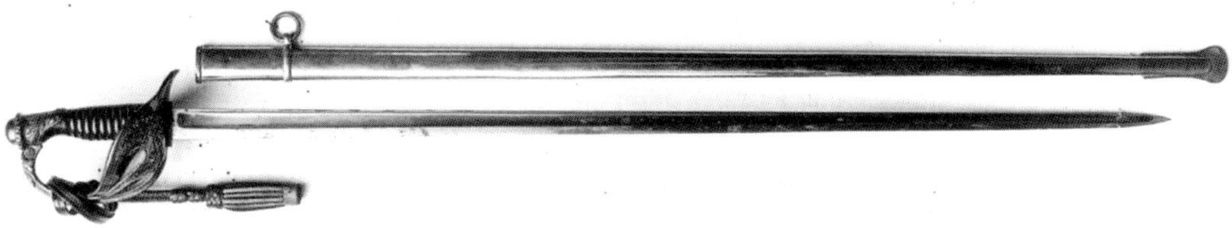

Figure 57. French Cavalry Sword Model 1896, the new French sword that Patton often referred to; it would be reissued in 1927. (*Courtesy of the Smithsonian Institution, Museum of American History*)

that the US saber of the time had all of the deficiencies of a curved sword and none of its benefits, it was difficult to point and thrust and at the same time was not a good cutting weapon. He would become more convinced of these ideals during his training in France after his participation at the V Olympic Games.

After participating in the 1912 Olympics, Patton honed his fencing skills at the French Army's cavalry school at Saumur under the tutelage of Adjutant Clery, then considered to be the foremost military fencing instructor in France. Clery exerted a great influence on Patton regarding swordsmanship and the proper use of the sword by cavalry. At Saumur, Patton fortified the belief he already held, with the sword the point was superior to the edge, and developed the outline he would use to write the new cavalry saber manual for the US Army, adopted shortly after the approval of his design for the cavalry saber.

In addition to the papers and reports Patton wrote advocating a straight sword for the cavalry, he also placed on display at the US Army staff headquarters in

Figure 58. This image from Saumur shows a French cavalryman giving point in much the same way that Patton would define in his writings on the use of the sword by cavalry. Note that in this image and in Figure 28, the thrust is delivered with the sword guard up, allowing greater reach and protection behind the horse's head and neck. (*Courtesy of the Library of Congress, Manuscript Division, Patton Papers, Box 74*)

Washington DC a relic from the Battle of Waterloo – a French cuirassier's sword, which would feature in the development of the US Model 1913 Cavalry Saber.[8]

Patton's report on his participation at the Olympic Games was dated September 19, 1912; five days later the acting Chief of Ordnance, Lieutenant Colonel Thompson, wrote the following letter to the commanding officer at the Springfield Armory:

> WAR DEPARTMENT
> OFFICE OF THE CHIEF OF ORDNANCE
> WASHINGTON
>
> September 24, 1912
> From: The Acting Chief of Ordnance.
> To: The Commanding Officer, Springfield Armory
> Subject: Experimental sabers.
>
> 1. With the saber which is being shipped you by express as enclosure 1 of 0.0. file 26791-0/2452, there is also being shipped you a model of the saber recommended for adoption by the Cavalry Equipment Board, an old French saber, the property of Lieutenant Patton, 15th Cavalry, and an English saber.
> 2. In addition to the carrying out of the instructions contained in the 4th endorsement, 0.0 26791-0/2452, it is desired that a new saber be manufactured under the following specifications:
>
> Blade: To be of the same length, shape, cross section and balance as the old French blade except that next to the guard the blade is to be left full size as in the experimental blade modeled after the French blade. The tang is to be of the same length as that on experimental saber with French blade and should be adapted to the gripe.
>
> Guard: Two guards, to be interchangeable, are to be manufactured for this saber. Both to be of the same material and of same gauge as that in the parabolic guard on the experimental saber with French blade. The edges are to be rolled in the same direction as on the Cavalry Equipment Board saber but are to be crimped as the edges are on the parabolic guard.
>
> No. 1 guard is to be modelled after that on the Cavalry equipment Board saber.

No. 2 guard is to be the same as No. 1 except that the distance of the edge of the guard from the gripe at the point A scratched on the guard is to be .375 inch greater than with the No. 1 guard.

Gripe: The gripe should be similar to that used on the experimental saber, drawing 20-2-83. The wire used in wrapping the gripe should be of copper and made up of two plain wires and a twisted pair as used on the saber recommended for adoption by the Cavalry Equipment Board and on some of the experimental sabers, shown on drawing 20-2-63. The gripe should have approximately the same curve and 'drop' as in the English Saber from the guard for 6.25 inches (approximately the length of tang now on experimental saber with French blade).

I. March 30th:
1. 2:00 p. m. **Exhibition of Mounted Swordsmanship.**
(a) Attacking dummies.
(b) Mounted combat.

The Instructor and Assistant Instructor in **Swordsmanship illustrating protection afforded by the hilt, from a cut at the head, in the position of charge saber.**

Figure 59. This image from a yearbook from the cavalry school at Fort Riley clearly shows how Patton carried over the influence of the French cavalry school to cavalry training in the US Army. (*Courtesy of the Library of Congress, Patton Papers, Box 75*)

The balance of the saber, assembled, should not [be] more than three inches from the guard and should approximate as closely as possible the balance in the old French blade.

3. The cost of this work is to be charged to your allotment for contingencies under the appropriation MANUFACTURE OF ARMS, 1912 AND 1913.

JNO. T. THOMPSON[9]

(Colonel Thompson would become a household name when later at Remington he developed the Thompson submachine gun.)[10]

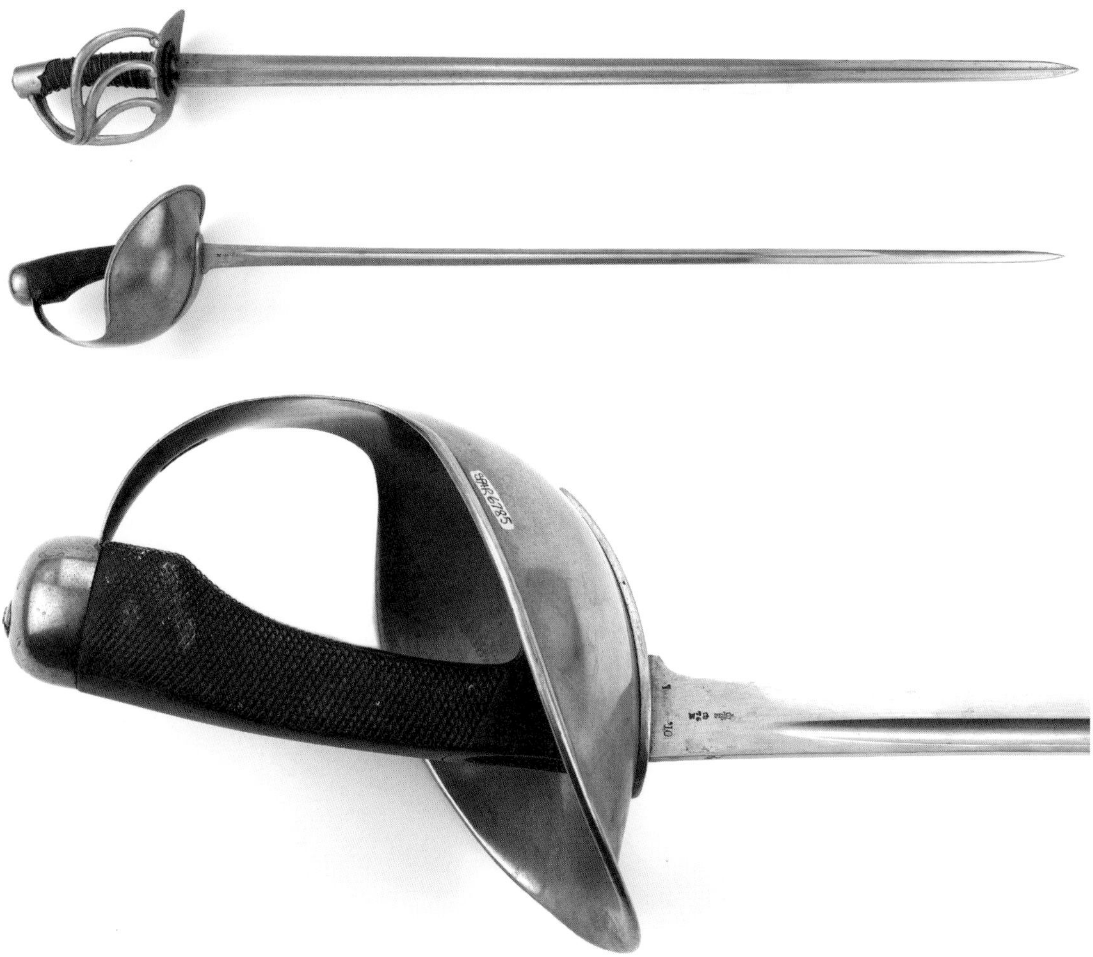

Figure 60. Top: the French cuirassier's sword that Patton had purchased in France and put on display at Army HQ, SPAR 6739; bottom: the British Model 1908 Cavalry Sword, SPAR 6785 sent to the Springfield Armory. (*Courtesy of the Springfield Armory Museum*)

The museum at the Springfield Armory historical site holds in its collection the French cuirassier's sword and the British Model 1908 Cavalry Sword sent to the armory as models for the new cavalry sword.

Based on the letter, Patton's ideas had already begun circulating at the headquarters and at least one quick mock-up had already been completed. The first experimental sword was made with a blade similar to a cuirassier's sword, a parabolic guard, and a grip similar to the 1905 experimental sword. Some others would be made with a guard similar to the 1911 experimental saber, the grip from the 1906 experimental saber and a blade based on the cuirassier's sword Patton provided. Patton himself owned at least one such sword. (See Figure 65 below.)

By January 1913, several changes were made to the sword, gone was the grip from the 1905 experimental sword; the tang and grip would be replaced by having the grip attached to the blade, like the handle of a kitchen knife or the popular US bolo knife. The blade would cease to resemble a cuirassier's sword and take on the appearance of the Swedish Model 1893 with a chisel point. These changes are described in a telegram from the new acting Chief of Ordnance Rogers Birnie to the commanding officer of the Springfield Armory:[11]

WAR DEPARTMENT TELEGRAM
OFFICIAL BUSINESS
WASHINGTON.

January 11, 1913
Commanding Officer,
Springfield Armory
Springfield Mass.

Prepare to manufacture two complete sabers, similar in design to experimental saber forwarded with twenty-six seven ninety-one 0 twenty-four eighty-two with following changes:

Blade on and one half inches longer and stiffer;
Metal to run through grip, similar to bolo;
Blade to have chisel point;
Guard to be made of stiffer metal.

Rock Island asked if sufficient metal on hand to manufacture guards, which could be furnished you. Full instructions and other minor changes by letter.

Birnie
Acting Chief of Ordnance.

Patton visited the Springfield Armory with other officers representing Army Chief of Staff Leonard Wood, and he made further refinements to the model provided by the highly skilled munitions workers. Patton improved the design of the grip and lightened the blade.[12]

By early February, the chief of the Springfield Armory, Lieutenant Colonel William Pierce informed the Ordnance Department that the armory had produced two sample swords and enumerated the equipment order and cost for producing the swords.[13]

Also, by early February, it became public that the armory had narrowed the choice for a new sword down to two similar designs. On February 13, 1913, the Springfield newspaper, the *Springfield Republican*, reported the progress of the new sword:

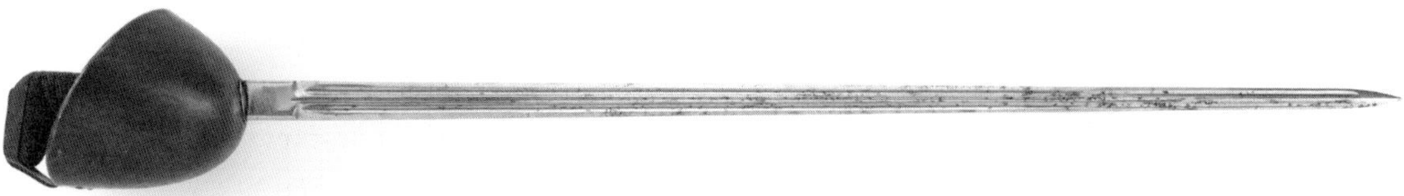

Figure 61. Experimental sword with 1906 experimental grip, parabolic guard and French-style blade similar to those carried by cuirassiers, SPAR 6707.

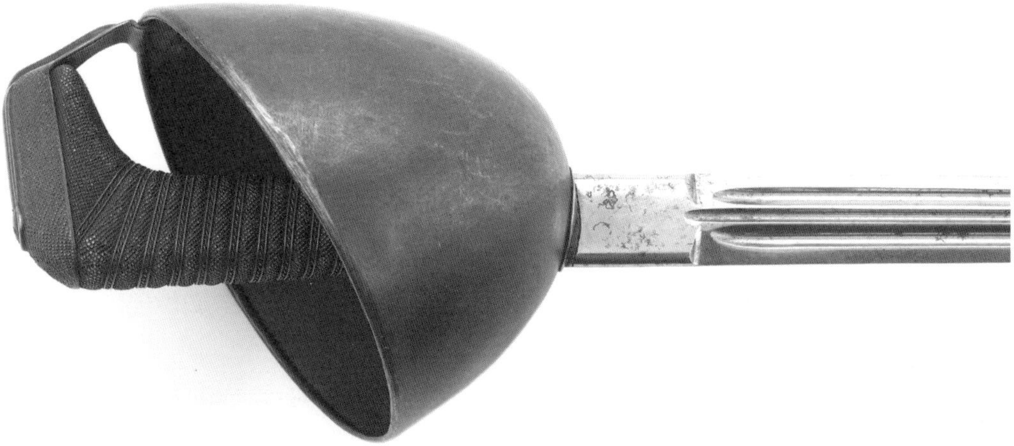

Figure 62. A close-up of the parabolic guard on the 1906 experimental saber.

NEW SWORD FOR THE ARMY.
Choice is Narrowed Down to Two
Models That Have Been Sent to
Washington

The local armory is about to begin the manufacture of a new type of saber for the United States cavalry. For months the weaponsmakers on the Hill have been turning out sample swords and sending them to Washington for approval and only last week a pair of sabers were packed and shipped which seemed to meet all the requirements. Doubtless the one will be chosen to supplant the old type of sword which has been in use in the cavalry since the civil war. Col. W.S. Peirce expects soon to learn from the ordnance department headquarters, of which of the swords is adopted, and then it is likely the local workmen will begin turning out the new weapons. Col Peirce said last week that the new branch of manufacture will not bring about any wholesale changes at the armory for no new men will be taken on the force, and only slight changes will be necessary in the machinery. The number of swords used in the cavalry is never large, and the job of equipping the horsemen will not be a huge one. There are now 15 regiments of cavalry in the regular army each averaging 800 men in times of peace and these with the officers of the other branches of the service are the only men to be equipped with the new weapon.

The local armory has not made swords for more than a dozen years. The demand from the army has always been slight and the type of sword used has never changed since the civil war. Some months ago the ordnance department was called upon to make a more modern saber and at once the experimentation began. Swords in profusion have been made at the local armory, but the choice has now narrowed down to the two which were lately sent to Washington. There is little difference between them. The Army and Navy Journal in

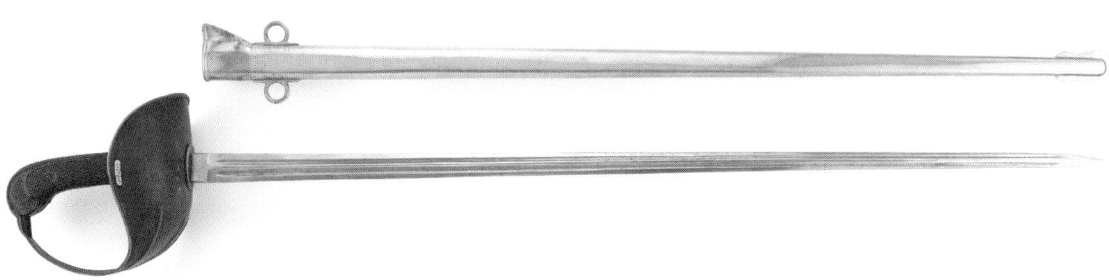

Figure 63. The prototype with final version of grip and guard but with a French-style blade and metal scabbard similar to what would be adopted for officers, SPAR 6863. (*Courtesy Springfield Armory Museum*)

an article in a recent number about the swords says: 'The only difference between the two is in the character of the grip. One of the swords has a thumb depression on the upper side near the guard and a stop on the underside for the index finger. The handles of both swords have a dropping knob which makes the sword lie in the hand in such a manner that its point will be practically on a straight line with the arm. As slight as the difference may appear in the grip of the two swords, there is quite a difference of opinion among swordsmen as to which is the better design. It is contended by some that the handle of a sword should be so shaped that the hand will slip when the point of the sword comes in contact with firm substance. In experiments conducted by the British army, where thrusts were made into beef carcasses, it is reported that the thrust with swords with thumb depressions and finger guard actually broke the arm of soldiers because the hand refused to give. This happened it is stated when mounted soldiers dashed with great force against the carcasses. While this might be an extraordinary occurrence, it is being urged that the grip should be so fashioned as to allow the hand to slip slightly in making the thrust.

Some of the army officers go so far as to claim that even the dropping knob of both handles is a dangerous grip. While primarily both the models are of a thrusting sword, the arm is to be so constructed that it will have all the slashing qualities of the saber now in use in the army. It will be somewhat heavier than the first model submitted to the general staff by the ordnance department. One officer in describing the new sword said, "It is to be a good big 'he' saber." It will be a straight sword with a ribbed blade about 35 inches in length with a chiseled point which has been shown to have great penetrating power, especially through cloth, canvas and goods of the heaviest texture. The blade is to be double-edged, and the tang will extend through the handle into the knob. The handle is to be covered on two sides with walnut. The wooden portion of the handle and the steel is to be checkered. The sword partakes of the characteristics of both the latest French and English models. It resembles the French service sword in that it has a straight blade and in the English in its strength and striking power in slashing. The point of gravity of the new sword will be about two inches from the guard. The guard is practically the same that was recommended by the cavalry equipment board, furnishing as much protection for the hand as possible and at the same time giving the hand freedom of action. The new sword which is being developed by the ordnance department will not be issued to infantry officers, according to the present plans of the war department. It will be exclusively for the mounted service and infantry officers will not be required to secure new swords when it is adopted.'[14]

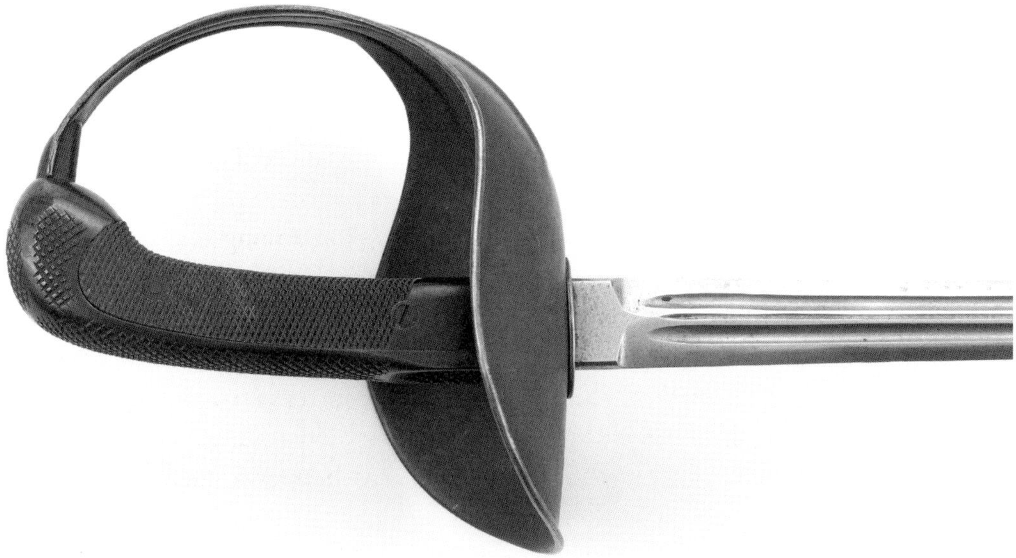

Figure 64. A close-up of the grip and guard. In January 1913, the *Army and Navy Journal* reported that the army was going to select a straight sword for the cavalry, based on a French sword brought back from France by Lieutenant Patton on display at the army staff headquarters. The issue contained an anonymous article about the importance of the use of the sword point by cavalry; Patton was the author. Besides eloquently arguing for the adoption of a straight cavalry sword in his report on the Olympics and in military publications, Patton completed a great deal of staff work for his boss Leonard Wood, the Army Chief of Staff.

The final design of the handle for the Model 1913 was similar to the grip on the final version of the handle on the last 1911 experimental sword. It had a checkered backstrap and thumb depression. The grip was mostly in line with the blade, and not set so that the point was online with the hand extended in the normal way, as Patton planned for the new sword to be pointed in the French fashion, with the knuckles of the sword hand up and the fingernails pointing down.

An article appeared in the *Springfield Republican* on February 18, 1913, describing Patton's visit to the Springfield Armory to supervise the final phases of production:

> HERE TO INSPECT NEW SABER – Lieut. George Patton Confers With Col. Peirce About Weapon to be Made at Local Armory.
>
> The new army saber which will probably be manufactured here occupied the attention of the armory officials yesterday as it will to-day. Lieut. George Patton, representing Gen. Leonard Wood, commander-in-chief of the United States army, came to Springfield yesterday morning and spent the

Figure 65. The prototype cavalry saber carried by George S. Patton, which with some alterations would become the Cavalry Saber Model 1913. Note that the blade is straight and double-edged, that the guard is very similar to the guard adopted on the Model 1913 Saber, but that the grip is the same as on the experimental Model 1906. Patton would improve the grip making it checkered and with a stronger kitchen-knife-style hilt, riveting the handle to the tang for strength. (*Courtesy of the General George S. Patton Museum, Knoxville, Tennessee*)

> day consulting with Col. W.S. Peirce of the armory about the design of the new weapon. For months the army experts have been experimenting with samples of the new sword made here after objection was overcome until now only one or two changes will be necessary. The saber is almost ready to begin its usefulness. Lieut. Patton brought the word from Gen. Wood that with small changes in the design of the handle and a lightening by 10 ounces the sword in its present style will probably be accepted. The latest model turned out by the local workmen weighs three pounds and eight ounces and will be reduced in weight to two pounds and 14 ounces. In design it is a combination of the approved French and English swords with a few ideas by American experts. It has a straight double-edged blade with a keen point and is intended primarily as a thrusting weapon. Lieut. Patton will stay in the city until the armory mechanics succeed in changing the weapon so that it will meet the expectations of the commander-in-chief of the army and the other men who have been developing the saber. Then he will return to Washington with the latest model and after that the making of the new sword in the local armory will no doubt begin. It will not necessitate the increase of armory equipment and the present force of mechanics will be able to handle the demand.[15]

On February 23, 1913, Wood sent a memorandum to the Chief of Ordnance ordering the production of 20,000 sabers based on Patton's design, and Patton was soon sent to the armory to work on finalizing the development of the new sword.[16]

The order for the production of the swords was reported in the *Springfield Union* on March 3, 1913:

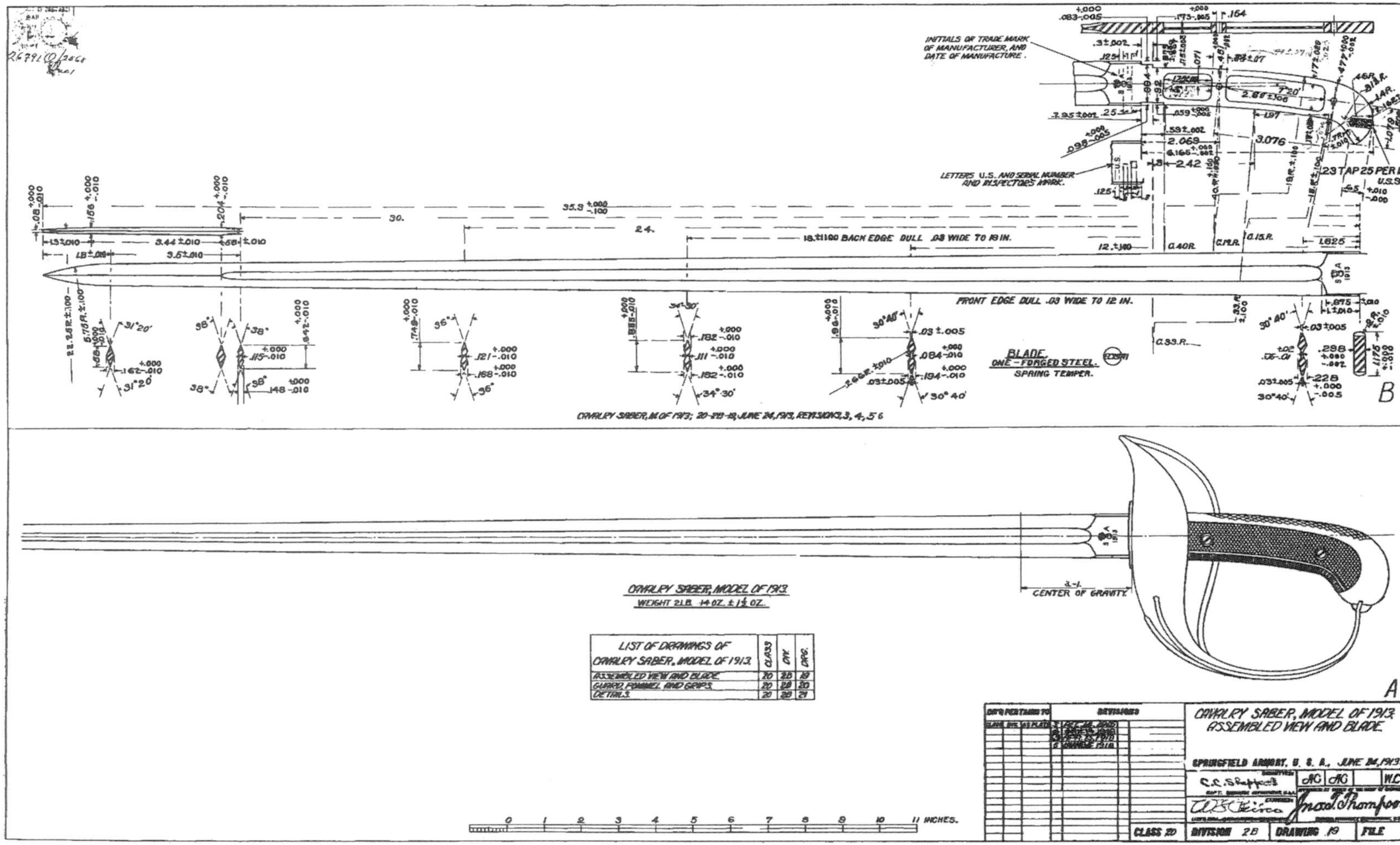

Figure 66. First of the Ordnance Department's technical drawings of the final version of the US Cavalry Sword and Scabbard Model 1913. (*Courtesy of the Rock Island Arsenal Museum Resource Center*)

ARMORY RECEIVES ORDER FOR SWORDS

20,000 of New Type of Cavalry Arm Will Be Turned out in Springfield

FINEST STEEL USED

First Order of Its Kind to Be Given Here: Enough to Equip Entire Force

The model of cavalry sword recently designed at the United States armory has been approved by the secretary of war and an order has been received by Col. William S. Peirce, commandant of the armory, for the manufacture of 20,000 swords. This order, according to Col. Peirce, will be sufficient to equip the entire cavalry division of the United States army.

The manufacture of the swords will not mean any addition to the regular working force of the armory, Col. Peirce said, but that all of the work will be done by the present staff of workmen. The new work will make up in a measure the difference in the rifle output for the coming year. This year's orders for rifles will be somewhat smaller than last year because of the large number of arms in storage in the different arsenals throughout the country.

Several new milling machines will have to be obtained before the manufacture of swords will be begun. These machines are equipped with tables especially adapted to the milling of the sword blades and it is expected that they will be received and set up in a few weeks. The work will be started as soon as the machines are set up and it is hoped to complete the order by Jan. 1, 1914, although it is possible that the work will be completed before that time.

Whether the manufacture of swords will be made permanent here, Col. Peirce could not say. He said that the manufacture of the swords is not such an essential part of ordinance work as the manufacture of rifles as a much smaller number are in use. It is probable, he said, that whatever additional swords are made above the original order of 20,000 will be manufactured in the local armory.

This will be the first time that cavalry swords will have been manufactured at the armory. A large number of officers swords have been made here in the past, but no cavalry swords have been turned out. The new sword is much different from the saber type of sword now used by the cavalry troops, and is the result of experiments in the manufacture of swords which has been going on in the armory for several months. During the course of the experiments three or more types were turned out, and these were submitted to the chief of ordnance and later to the secretary of war for consideration.

> The new sword is believed to be one of the best types of cavalry sword turned out from any armory in the world. According to experts who have examined the type of steel used in manufacture of swords and the fine finishing process given them in the local armory they equal or excel any sword made in the world. The new sword will be made of the best steel obtainable, and the work of finishing will bring it down to the finest temper.[17]

In April, Colonel Rogers Birnie, acting Chief of Ordnance, wrote to the Adjutant General extending his appreciation for Patton's assistance in the final adaptations for production of the Model 1913, noting that Patton's experience and skill as a swordsman were invaluable.[18]

As described in the War Department Annual Reports, 1913: 'The new sword has a double-edged straight blade, 35 inches in length, the guard and grip being modeled after the design recommended by the Cavalry Equipment Board.' The guard on the saber was sheet steel with three corrugations to improve strength, the only guard of its kind on an American sword; the grip was checkered steel with checkered hardened rubber inserts on the sides. One detail about the grip that the War Department omitted to mention was that it had an almost rectangular cross section, making it much more secure in the hand and less likely to inadvertently turn in the hand as would be the case with a more rounded grip. In addition, like many of the best European sabers, it had an indentation on the backstrap for the thumb. (It should be noted that the final version of the 1911 experimental saber had a similar guard, with three corrugations, and the Model 1913 benefited from the guard and backstrap developments of the 1911 experimental saber.) The hilt was in line with the blade, which may seem un-ergonomic, except when you consider that Patton instructed cavalrymen charging with the saber to lean forward in the saddle to protect themselves with the horse's neck and extend and rotate the sword arm so that the thumb was below and the knuckles above. (See Figure 59 above.)

The report further describes an officers' model with a nickel-plated scabbard for dress and garrison use, but a web-covered hickory scabbard for field use by both officers and enlisted men.[19] The War Department reported in 1914 the satisfactory manufacture of the new weapons; the nomenclature was changed to 'saber and scabbard model of 1913'. Almost 25,000 were manufactured by the time of the 1914 report.[20]

Patton was well pleased with the final result. In 1916, just before serving in Mexico, Patton watched a parade of US cavalry all equipped with his sword: 'My eyes filled with tears … the nearest similar feeling I can remember is on the occasion of a certain very noisy Opera music.'[21] Patton was in Mexico because on

March 9, 1916, Pancho Villa led a foolhardy raid into the United States, killing civilians in Columbus, New Mexico, setting off the Wilsonian response of the Punitive Expedition.

Patton was on the staff of his mentor, General John J. Pershing, when he learned the value of airplanes and motorized vehicles in war. In his diary on April 8, 1916, he noted that three airplanes were able to scout 700 square miles. And in one of those prophetic episodes of history, he led the very first motorized attack by the US Army on an enemy outside the United States.[22]

On the way to purchase corn in San Miguel, Patton led his force of fifteen men in three automobiles to investigate several ranches, as he suspected they were sheltering supporters of Pancho Villa. At the Rubio ranch, about 10 miles south of San Miguel, Patton and his men leaped from their automobiles into action against mounted enemies. During the short skirmish, Patton and his men killed one of Villa's leaders, Colonel Julio Cardenes, along with Captain Isador Lopez and Private Juan Garza.[23]

New York Times newspaper reporter Frank B. Elser publicized this new form of warfare, recognizing its importance and, from a US standpoint, its novelty.[24] But he could not at that time have imagined the tremendous role Patton would play in modern motorized and mechanized warfare.

With the success of the gunfight near San Miguel behind him, Patton felt confident enough to argue his thoughts about the saber being essential among the equipment of the cavalry with General Pershing, who grudgingly agreed to a published article to that effect.[25]

Upon returning from the Punitive Expedition, some cavalry officers expressed, in the *Cavalry Journal*, that, unused in Mexico, the saber was an unneeded encumbrance. Patton came to his sword's defense, stating that not a single cannon shot was fired in Mexico, but no one advocates abolishing artillery.[26]

Patton was clearly entitled to the elation and pride he felt for his creation. The last cavalry sword of the US Army was a cavalry-only weapon to be attached to the saddle, not worn by the soldier, the way that an army officer who was not on cavalry duty would wear the 1902 saber.

Patton authored the manual *Saber Exercise, 1914*, which begins with a brief introduction by Major General Leonard Wood, Chief of Staff: 'The following Saber Exercise prepared by Second Lieutenant George S. Patton Jr. Fifteenth Cavalry. Master of the Sword at the Mounted Service School and revised by the Cavalry Board is approved and issued for the information and government of the Regular Army and the Organized Militia of the United States.'[27]

In the manual, Patton describes his cavalry sword: 'The blade is divided into the forte, the 18 inches nearest the hilt; and the point, the rest of the blade. The saber;

Model 1913 is two-edged. All the front edge and half the back edge is sharp so that it may be more easily withdrawn from a body and also on rare occasions used to cut.'[28] Writing in 1930, Patton no longer held the view that the double edge was necessary:

> In 1912, when I was working on the saber I designed … Gen. H. T. Allen and E. St. J. Greble were on the G.S. [General Staff] and insisted that I make a double edged sword so it would pull out easier (this was the theory as they had never tried it in bodies of pigs as I had – the single edge pulls out equally well).[29]

It should be noted that the French Model 1896 Cavalry Sword was single-edged for much of the length of the blade; this would have been the sword in use when Patton studied at Saumur.

Throughout most of his life, Patton would continue to see the value of mounted cavalry and the cavalry sword. In fall 1919, responding to the French Cavalry Officers' Board, Patton wrote that the saber should be retained to stimulate morale and dash; 'it is worth its weight.'[30]

Even as late as 1941, when Patton was putting a tank brigade through its paces at Fort Benning, Georgia, he still expressed his views regarding the value of cavalry and the cavalry sword. When the Cavalry Board asked his opinion about whether cavalry men should still be armed with the saber, Patton responded with a forceful affirmation: 'A cold steel weapon is not only desirable, but vitally necessary.'[31] And he continued to explain that a charging infantryman is more easily checked than a charging horse and rider. Using examples from history, he described nations whose cavalry abandoned the sword only to take it up again. He closed by stating that what made the sword valuable was not just its ability to wound or kill but the fear it inspires in the enemy: 'People on foot fear cavalry, that's why municipalities retain mounted police.'[32]

Chapter 10

The Patton Cavalry Saber in War and Peace

Although Patton would depart for France as a member of Pershing's staff and eventually become one of the leaders of the American tank force during the First World War, the US cavalry which took an active role in the defeat of the Germans did so without Patton, for fate had a different plan for his future. Patton had learned the value of new advances in technology during the Punitive Expedition in Mexico, but even before that, he had an interest in technology. Patton owned an automobile as early as 1910, and his letters to his father of the time showed a thorough knowledge of the automobiles of the period.[1] Now, only very old people remember a time when foreign-made automobiles made up

Figure 67. Lieutenant Colonel George S. Patton in front of a Renault tank in France during the First World War. (*Signal Corp image courtesy of the United States Archives*)

only a small percentage of the market and only those who are centenarians or older remember a time before the previous domination of the market by the big three – General Motors, Ford (Lincoln Mercury), and Dodge-Chrysler-Plymouth. By the start of US involvement in the First World War, there were numerous automobile manufacturers, whose names, such as Winton, Mercer, and Stutz, now have largely been forgotten. Patton's familiarity with automobiles would prove useful during the next phase of his life.

Within a month of the United States' entrance into the First World War, Patton was attached to the staff of the general commanding all forces, his mentor from the Punitive Expedition in Mexico, John J. Pershing; by June, they were in France.[2] And through the summer and into the fall, Patton held various roles on Pershing's staff, including head of the motor pool. During that time, it was unclear to Patton what the future would hold, and his attachment to the cavalry was strong, as shown by an anecdote in his diary about meeting General Haig of the British Army, with whom he discusses his saber design, as Haig was in his heart and origins a cavalryman.

However by September, he had become interested in tanks, and by October, he accepted a promotion to major to head the newly established US Army Tank School at Langres, France.*

The tank was an invention born of the necessity of the First World War and the trench warfare that defined it. Armored cars were not up to the task, being unable to cross trenches or keep from getting stuck in the mud. The British were the first to see the military potential of tracked vehicles, which prior to the war had been designed for converting undeveloped land to farm use, and some tractors used in Britain were of US design. The British were the first to build and then employ these armed and armored tracked vehicles in battle in September 1916. The French would first employ their tanks the following April.[3] These early vehicles were held together with enormous bolts and rivets, and because of their similarity in appearance to large oil tanks, they were called 'tanks,' the name they are known by to this day.

Patton had an innate understanding of and interest in things mechanical, and while he and his technical assistant, Lieutenant Braine, toured the French Renault tank works, they made suggestions that the French add a bulkhead between the gun room from the engine and add a self-starter to the engine; the French adopted both suggestions.[4] Although Patton pushed hard for US tank production, the first two US tanks arrived after the war was over. The US Army would fight using French tanks.

---

* Patton was not actually promoted to major until December 1915; he would be promoted lieutenant colonel at the end of March. Martin Blumenson, *The Patton Papers, Vol. 1: 1885–1940* (Houghton Mifflin, Boston, 1974), pp. 429, 486, 513.

While organizing the Tank School, Patton also determined the specifications for American tanks and thought out the organization of tank units. By February, he had his first class of 200 men.[5]

By August 1918, Patton had so well organized the school that he was able to turn it over to a subordinate and move on to the planning of engaging tanks in battle.

The St Mihiel Offensive

During the St Mihiel offensive, Patton commanded a tank brigade comprising two battalions – the 326 and the 327.[6] Patton's tank brigade would support the 42nd Division in the Allied effort to push north starting on an east–west line from St Mihiel on the River Meuse in the west to Pont-à-Mousson in the east on the Moselle. The St Mihiel salient had been an objective since 1915, as it cut major rail lines between Paris and Nancy; this would be the first major US effort of the war.[7]

Two days into the battle, Patton and Douglas MacArthur, commander of the 84th Infantry Brigade, met on the field of battle during a shelling and at one point, uncharacteristically, Patton flinched, MacArthur retorting dryly, 'Don't worry major, you never hear the one that gets you.'[8]

The town of Essey was about a third of the way to the ultimate objectives and Patton's tanks were on their way to take it when French soldiers warned the tankers that the bridge was too dangerous; Patton exhibited his bravery leading his tanks, on foot, into the town. At the next town, Pannes, about halfway to their objective, Patton sat on top of a tank to show his disdain for enemy fire, leading the infantry into the town. By September 15 the Allies had realized their objective and Pershing officially recognized the contribution made by the tank brigade.

During the subsequent Meuse-Argonne offensive, when his tanks were held up by trenches they couldn't navigate, Patton, ignoring German machine-gun fire, led his men in filling in the trenches with shovels. Once his tanks were past the obstacle, he followed on foot and soon afterwards he was wounded by rifle fire. He was hospitalized and by the time he fully recovered, the war was over.[9]

US Cavalry in the First World War

When the Second Cavalry under Colonel O.P.M. Hazzard arrived in France in April 1917, they were equipped with rifles in 30-06, Colt semi-automatic pistols in .45 caliber, gas masks, and Model 1913 Sabers.[10] The Second Cavalry, armed with their pistols, rifles, and sabers, was the only American cavalry unit to take part in a major offensive, not as dismounted cavalry, but mounted. They contributed to the success of the St Mihiel attack.[11] And although there is no record of the saber being

Figure 68. US Second Cavalry soon after arriving in France, with their Model 1913 cavalry sabers in hand. (*Signal Corps*)

used against German soldiers, it cannot be proven that no US trooper used his saber in action against the enemy.

After the war, Lieutenant Colonel Charles A. Romeyn, an officer with the chief of staff, wrote: 'I am confident certain troops of the Second Cavalry were in mounted action at St. Mihiel and perhaps elsewhere. Did they use the saber, or the

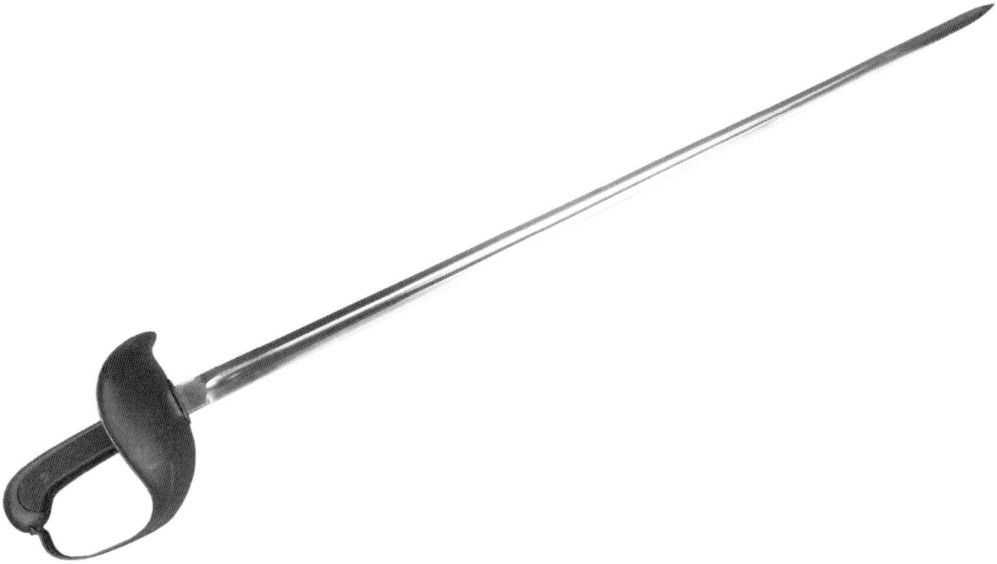

Figure 69. The first Cavalry Saber and Scabbard Model 1913, produced by the Springfield Armory. (*Courtesy of the National Museum of the US Army, Fort Belvoir, Virginia*)

pistol, or both? And what were the results? We have a case in our own Army. What happened?'[12]

Patton had not intended to stay in tanks; as he told his wife, Beatrice, he saw it as a specialty and specialists didn't get supreme command.[13] Patton returned to the cavalry after the war.

The Bonus March

Soon after the veterans returned from service in the First World War, some through the Veterans of Foreign Wars and the American Legion petitioned Congress for additional compensation for their service in the war. In 1924 a law was passed to

Figure 70. The Patton home at Christmas, with daughters Ruth Ellen and Bea, but note the Model 1913 Officers' Saber to the far left. (*Courtesy of the Library of Congress*)

provide veterans of the First World War with a monetary compensation in 1945, colloquially referred to as the 'bonus.'[14]

During the height of the Great Depression, some veterans sought an early payment of the bonus and in February 1932, thousands of veterans and hangers on marched on Washington. They hoped to sway Congress, but when the session ended for the summer without the passage of legislation many left but thousands remained, living in shacks in vacant lots and squatting in empty government buildings. At this same time Patton had just begun serving as the second in command of the Third Cavalry Regiment at Fort Myer in nearby Arlington, Virginia.[15]

During a scuffle between protesters and police on July 28 a policeman shot a protester and because of the ensuing disturbances, army units at Fort Washington and Fort Myer were called in. The cavalry from Fort Myer arrived first, taking up a position on the Ellipse, near the White House at 2.40 p.m. While they waited for the infantry from Fort Washington, Patton, on horseback, made a reconnaissance, riding down Pennsylvania Avenue as far as 3rd Street, returning with a full understanding of the situation. As Blumenson summarized: 'The clearing action started at 4:05 p.m. Cavalrymen wore steel helmets, carried gas masks and carbines, at the sling, and held drawn sabers.'[16]

As Patton described the action, the line of cavalry surged into the protesters, who threw bricks and were met with saber strokes (the flats of the blades). 'Two of us charged at a gallop and had some nice work at close range with the occupants of the truck, most of whom would not sit down for some days [afterwards].'[17]

To Patton, it was a distasteful business, but his duty to protect the Federal city was clear. The noble saber had been set to an ignoble but necessary task.

Chapter 11

The End of the Saber

As the world would approach the commencement of the Second World War, the US Army would undergo changes, particularly in the role of cavalry. Although the army discontinued the use of the saber, in April 1934, cavalry was still an important arm of the US army.[1] But as the United States was slowly readying itself for the coming challenge, the army engaged in reorganizations.

The complete replacement of horse cavalry by motorized and mechanized units was not a foregone conclusion in the late 1930s. As German and Japanese aggressions heightened, in early 1938, the War Department mandated two branches – cavalry and infantry – to develop mechanized forces.[2]

Summer 1938 was a significant one for Patton, he was promoted to full colonel and given the Fifth Cavalry Regiment to command at Fort Clark, Texas. At the

Figure 71. A US Army officer (right) during a review for foreign officers in Washington, DC during the 1920s, Note the Model 1913 Officers' Saber and Scabbard.

end of the war games that August, the Third Army commander concluded that horse cavalry had continued to show its value, no doubt because of Patton's excellent handling and understanding of the value of mobility.[3] After Fort Clark, Patton commanded the Third Cavalry at Fort Myer.

As the world braced for the oncoming war in summer 1939, Patton was involved in war games, again, this time the maneuvers were near Manassas, Virginia.[4]

After the German invasion of Poland on September 1, 1939, President Roosevelt raised the US Army strength to 227,000 men. At the same time, the army adopted triangular divisions – each division would have three regiments and each regiment, three battalions, instead of four and four, respectively. The change from rectangular divisions to triangular divisions was to aid mobility and flexibility.

During the war games in early spring 1940, the mechanized cavalry and motorized tank units overcame the horse cavalry units, coinciding with the success of the German blitzkrieg in France.[5] By the summer, Patton had accepted the command of an armored brigade at Fort Benning. Horse mounted cavalry would have no combat role in the US Army in Europe during the Second World War; the cavalry saber would not be part of the equipment of the American fighting man in that conflict. However, the Patton cavalry sword did have a legacy in the Second World War; thousands were ground down and altered to make fighting knives for US combat troops.

Patton, however, would go on to be indispensable to the Allied victory in the Second World War, because of his experience, ability, and unswerving dedication to the profession of arms. The last cavalry saber of the US Army, although it was not an important weapon, was an important example of Patton's thorough approach to his duty as a cavalry officer. As Aleksandra Rohde wrote in her introduction to writings by Patton about the sword: 'While his elegant sword would never win a major battle, the aggressive tactics and the warrior ethos he inspired … would endure and help to rescue Europe.'[6]

Ultimately, the nineteenth-century Chief of Ordnance, S.V. Bennet, may have been prescient to suggest the obsolescence of the sword. Colonel J.C. Kelton, the Assistant Adjutant General, agreed and went further to suggest that the sword was only necessary among armies on the continent of Europe because of its association with chivalry and the role of the knight as an officer over lower class soldiers, but because of the republican traditions of the United States, the sword had no place as a weapon or symbol of office.[7] However, as previously reported, the Secretary of War took no action and the sword remained.

Finally, as the sword was the last emblem of chivalry, Patton was the United States' last chivalrous knight, and so it is fitting that he should have designed the nation's last sword. The sword is often associated with European aristocracy, and Patton was an aristocrat. However, it should be remembered that 'aristo' is from the Greek word meaning 'the best,' and Patton was certainly the best of his kind. Although he was the descendant of southern aristocrats, the grandson of a confederate colonel, and the son of California's landed gentry, he never rested on his station or wealth. He understood his obligations under the rules of *noblesse oblige*. His duty was to protect those weaker than himself. To do that, he unswervingly sought improvement as an army officer, and bravely faced, indeed sought out, every danger and challenge.

Patton's leadership in the world wars was the basis of his enduring fame. The common soldiers were glad to be led by him. As Corporal Frank Sisson recollected in his memoirs: 'The thought of joining up with General Patton's Third Army left me with eager anticipation. From everything I heard, this was the general to serve under.'[8]

After taking a leading role in whipping the pre-war US Army into fighting shape, successfully leading US forces in North Africa, winning at Casablanca, in Tunisia and Sicily, exploding the breakthrough after Normandy, and saving the US Army at Bastogne, Patton led the US Third Army in its victorious campaign into Germany. A little more than seven months after Germany's surrender, Patton was killed in an automobile accident, the day before he was due to fly home.

The accident occurred on the way to hunt pheasant in Mannheim. A truck cut off Patton's staff car, he was thrown forward hitting his head on the ceiling because he was sitting perched on the edge of his seat to better see and be seen. The blow broke his neck, paralyzing him. In charge of his treatment was Colonel Glen R. Spurling, the chief neurological consultant for the army in Europe.* However, this outstanding doctor was unable to save General Patton, who died of an embolism on December 21 in hospital in Heidelberg, Germany.[9] He was buried among US soldiers in Luxembourg, heroes who gave the fullest measure in the Battle of the Bulge and with them lies the greatest warrior of the Second World War.

Some conspiracy theorists have suggested foul play, but they should consider the activities of his extremely active life and his numerous close calls with death. He was wounded in the First World War, by rifle fire and had numerous accidents throughout his life. During the Second World War, while inspecting a ruined tower near Naples, a salvo of German artillery rounds left shell craters 30ft off from where Patton and his companions had been standing; Patton had stopped to take a photograph, and then continued; had he not stopped when he did and then continued when he did, he and his companions would have been killed.[10]

On November 11, 1953 (Armistice Day and General Patton's birthday) the US Postal Service issued a 3-cent stamp commemorating General Patton's wartime

---

* Interestingly, at Walter Reed Medical Center in Washington DC, Spurling saved former Patton North Africa subordinate General Robinett from losing his leg.

service. As part of the commemoration the postal service summarized Patton's life and extolled his virtues, particularly his ability to transform his cavalry tactics into a winning formula for modern mechanized warfare, as well as the loyalty he instilled in his troops.

The Second World War US Fighting Knives Made from Patton Cavalry Sabers

At the start of the Second World War, there was a need to supply American soldiers with fighting knives, and companies like the San Antonio Iron Works of San Antonio, Texas, and the Robinson Company, in Glendale, California, converted

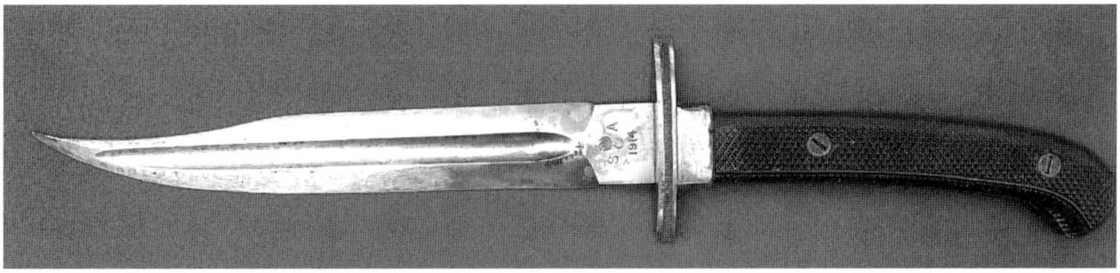

Figure 72. Fighting knife made from the grip, guard, and forte of a Model 1913 Cavalry Saber. This is the finest example the author has seen of one of these fighting knives. It is marked on the obverse: US (above the serial number 9042). On the reverse it is marked with the Springfield Armory mark: an 'S' and an 'A' separated by a flaming bomb, centered above the year of manufacture 1914.

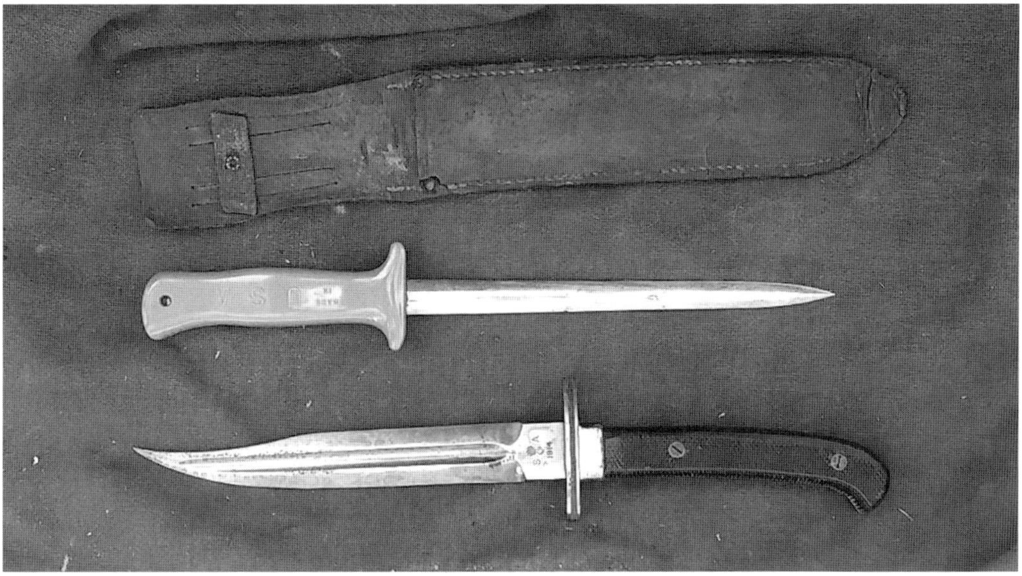

Figure 73. Two fighting knives made from Model 1913 Cavalry Sabers; the top knife was converted from a Patton saber by taking the tip-most section of the blade and imbedding it in a hard plastic handle; it is shown alongside the knife in Figure 72 for comparison.

US sabers into fighting knives. Although some earlier sabers were converted, primarily the Model 1913 Cavalry Sabers, the Patton swords were converted. The blades were basically divided into three parts. Those with the blade closest to the guard and hilt incorporated the original grip and cut down the guard (see Figure 72 above). These are the rarest and most valuable variants and when they appear on the market, disappear quickly and sell for well over $1,000, which is more than actual Patton Sabers sell for. The collector might wonder why this is so, but beyond scarcity, an analogy could be drawn from the antique watch world: a high-quality antique pocket watch can be found for a few hundred dollars, while a similar quality antique wristwatch may command a significantly higher price, because no one wears pocket watches, but people still wear wristwatches. Whereas it's unlikely that someone would carry one of these Bowie-Knife-like Second World War fighting knives as a hunting knife, the owner could.

The remaining middle section of the blades were also converted into fighting knives as were the points, usually embedded in a plastic handle, see the top knife in Figure 73 above, made by the Robinson Knife Company.

Chapter 12

Sword Production

The first Patton cavalry swords – they were officially designated as the Cavalry Saber Model 1913 by the Secretary of War – were produced at the Springfield Armory in Springfield, Massachusetts. During 1913–14, the armory produced a total of 24,799 of the new Patton cavalry sabers.[1]

During 1913, the Springfield Armory turned out almost 8,500 Model 1913 Sabers – the highest known serial number for a saber marked '1913' is 8140; in 1914, almost 27,800 were produced and another 5,300 in 1915.[2]

From July 31, 1917 to June 30, 1918, the Springfield Armory produced an additional 7,661 Model 1913 Sabers, before the entire focus of the armory was turned to solely producing the Model 1903 Infantry Rifle.[3]

Production of the Model 1913 Cavalry Sword and Scabbard was then turned over to the New Britain, Connecticut, company, Landers, Frary & Clark. The illustration, Figure 79 below, from a collection of advertisements from Connecticut companies, shows the various defense products produced by Landers, Frary & Clark to support the American effort in the First World War; the most prominent item in the illustration is the Patton saber.

It is estimated that Landers, Frary & Clark manufactured 54,000 of the Saber and Scabbard Model 1913, including all of the nickel-plated officers' swords and scabbards (see Figure 78 below).

As a result of the considerable production of the Model 1913 Cavalry Saber, they are not rare and can frequently be picked up at gun shows and antique shows at a reasonable cost. Although the price guide by Harvey Withers estimates $775, the author's example in this book – a 1913 Springfield Armory issue – was purchased at a gun show in March 2019 for $450. As recently as July 2021, the author saw some Model 1913s for sale, without scabbards, for $250. The officers' models, which are scarcer and nickel-plated, are somewhat more expensive.

It has come to the author's attention that reproductions of the Patton Cavalry Saber are being manufactured in India; buyers should beware and examine carefully before purchasing. Special attention should be given to the stampings on the ricasso (see Figures 74–8 below).

Figure 74. Hilt, guard, and ricasso of a model 1913 saber, obverse showing the letters 'US' and the serial number 1020, early in the production run. (*Author's Collection*)

Figure 75. Hilt, guard, and ricasso of a Model 1913 saber, reverse showing the Springfield Armory initials 'S' and 'A' separated by a flaming bomb and the year '1913'. (*Author's Collection*)

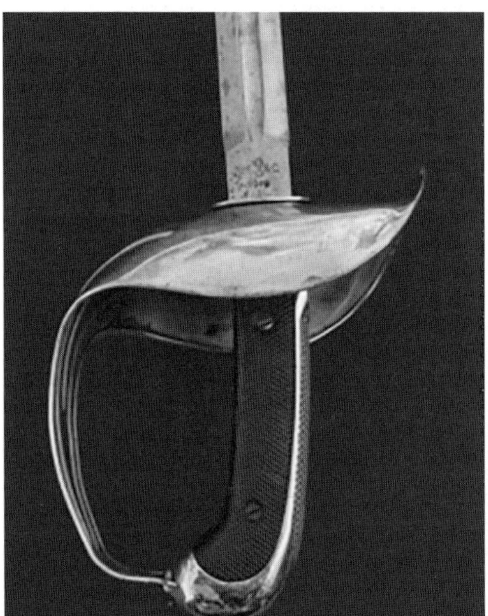

Figure 76. Hilt, guard, and ricasso of an officer's Model 1913. Note the markings 'LF' and '&C' separated by a flaming bomb above the year of production 1918. (*Fogel Collection*)

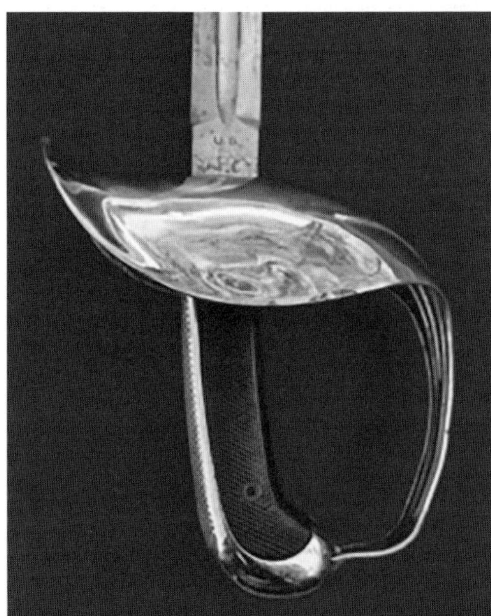

Figure 77. Hilt, guard, and ricasso of an officer's Model 1913. (*Fogel Collection*)

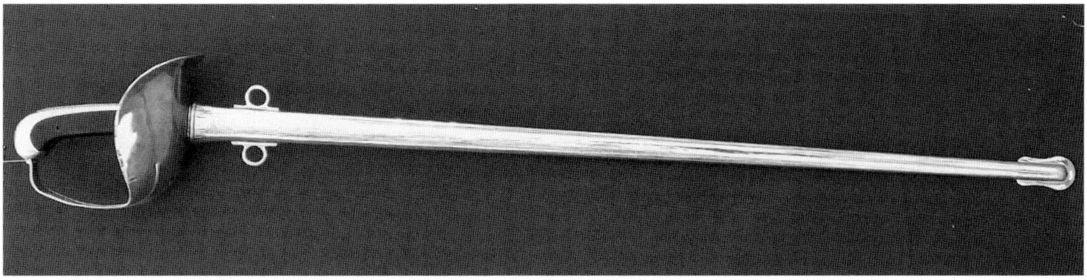

Figure 78. Model 1913 Officer's Cavalry Saber and Scabbard. (*Fogel Collection*)

The Smithsonian's Collection of Patton Cavalry Sabers

The Smithsonian Institution National Museum of American History has a large collection of swords, primarily US military and European military swords, however, its collection also contains swords from the Near East and Japan. There are five Patton Cavalry Sabers in the collection, and these are shown and described on the following pages. Also shown are two officers' models from private collections.

Figure 79. A page from the publication *New Britain's Welcome Home to Her Service Men* showing the wartime production of Connecticut-based company Landers, Frary & Clark; most prominent in the image is the Model 1913 Cavalry Saber. (*Courtesy of the University of Connecticut Library and the Connecticut State Library*)

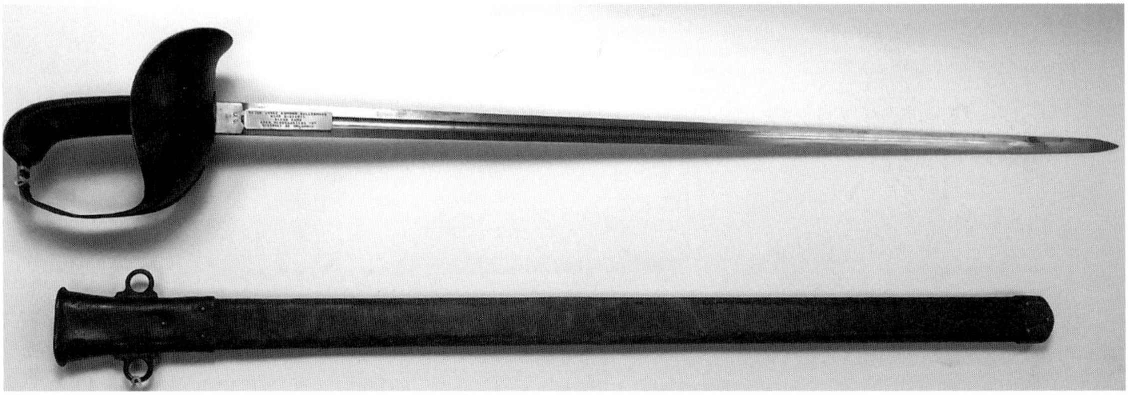

Figure 80. The sword with the small plaque near the guard should be captioned, 'Figure US Saber Model 1913 courtesy of the Smithsonian Institution, National Museum of American History.' This sword belonged to Major Josef Egmond Gellermann – later a professor at Georgetown University, in Washington, D.C.

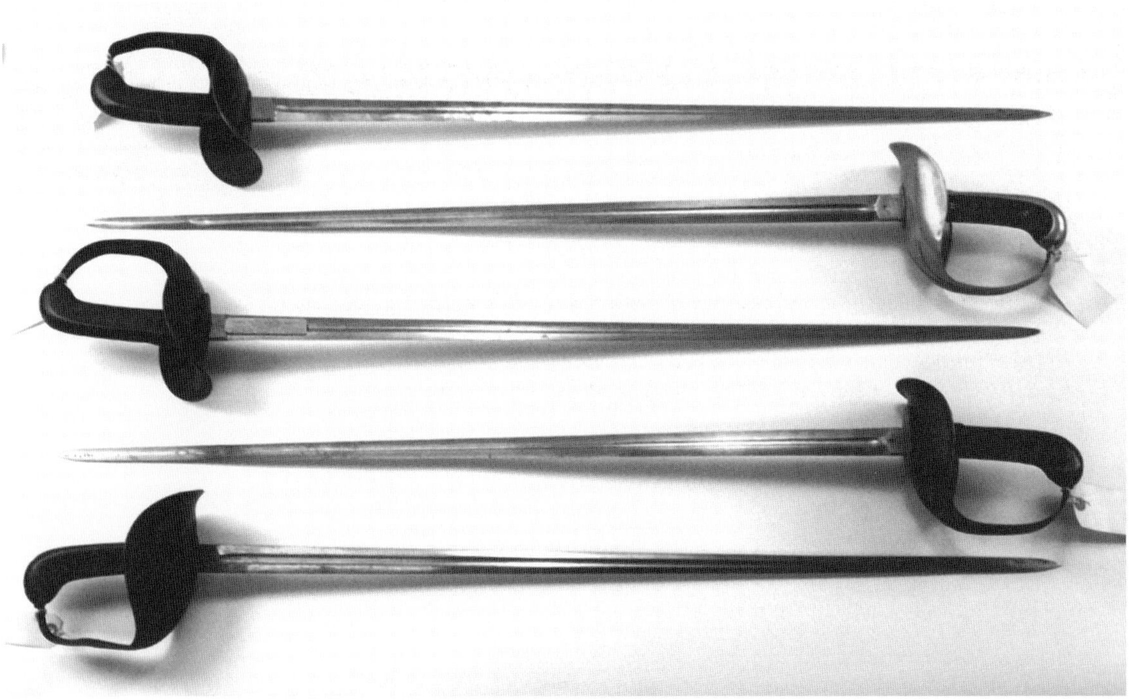

Figure 81. All of the Model 1913s in the Smithsonian Collection.

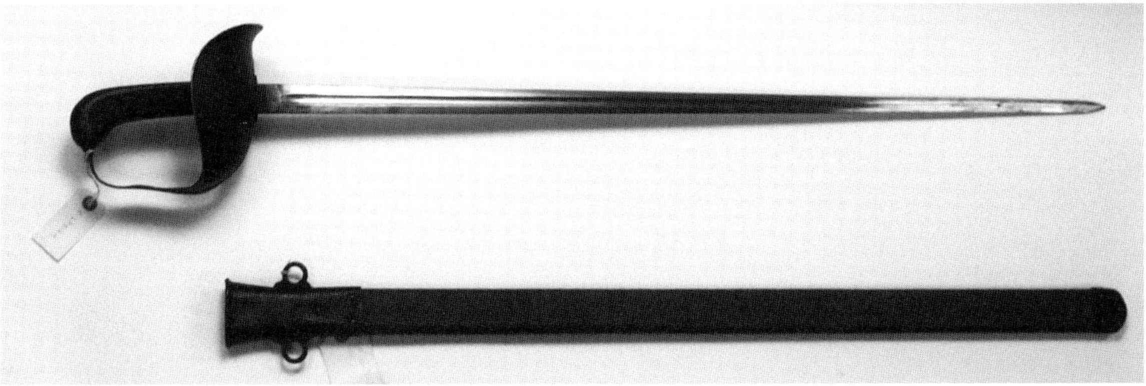

Figure 82. US Cavalry Saber Model 1913, without markings and scabbard. Notes: C.67812M was purchased from Norm Flayderman, one of the foremost authorities on US swords, in 1962, back when his shop was in Norwich, CT, for $20.53. (*Courtesy of the Smithsonian Institution, National Museum of American History*)

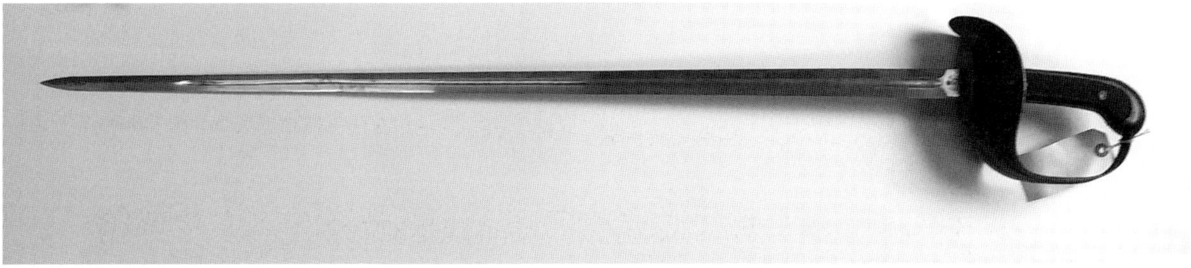

Figure 83. This is the sword described in *American and European Swords in the Historical Collections of the United States National Museum*, by Theodore Belote. Catalog number: 83513H18, which was collected by the Smithsonian before the National Museum of American History opened and was transferred to its current location in 1965; it was thus the first Model 1913 acquired by the museum.

Appendix 1

1913 Springfield Armory Technical Drawing, Model 1913 Cavalry Saber

The following six pages show the technical drawings used by the Springfield Armory to manufacture the Model 1913 Saber and Scabbard designed by then Lieutenant George S. Patton. The images are courtesy of the Museum Resource Center at the Rock Island Arsenal, Rock Island, Illinois.

The first illustration shows the overall design of the entire sword. The second illustrates the design of the guard and grip. (Note the three corrugations on the guard for added strength with minimal added weight, the diamond patterned grip to make it easier to handle, and the thumb depression, to make sure the cavalryman had a good hold in the proper position.) The third illustration shows the various ferrules. The fourth illustration shows the officers' scabbard, and the fifth shows details pertaining to that scabbard. The sixth illustration details the troopers' scabbard.

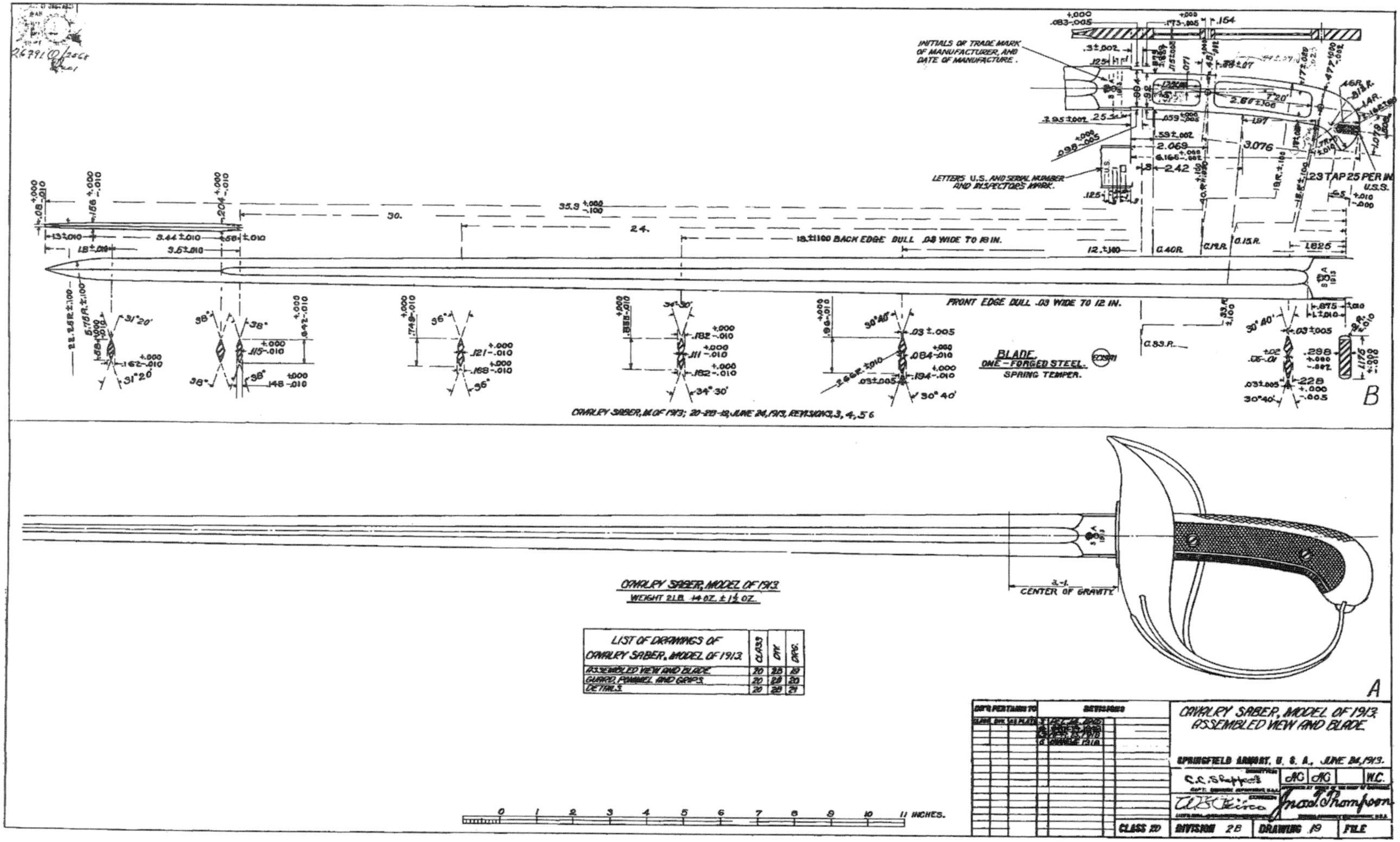

INITIALS OR TRADE MARK OF MANUFACTURER, AND DATE OF MANUFACTURE.
LETTERS U.S. AND SERIAL NUMBER AND INSPECTOR'S MARK.
23 TAP 25 PER IN. U.S.S.
18.±.100 BACK EDGE DULL .03 WIDE TO 18 IN.
FRONT EDGE DULL .03 WIDE TO 12 IN.
BLADE
ONE-FORGED STEEL.
SPRING TEMPER.
CAVALRY SABER, M. OF 1913; 20-28-19, JUNE 24, 1913, REVISIONS 3, 4, 5 6
B
CAVALRY SABER, MODEL OF 1913
WEIGHT 2 LB. 14 OZ. ± 1½ OZ.
CENTER OF GRAVITY
A
LIST OF DRAWINGS OF CAVALRY SABER, MODEL OF 1913.
CLASS
DIV.
DRG.
ASSEMBLED VIEW AND BLADE 20 28 19
GUARD, POMMEL AND GRIPS 20 28 20
DETAILS 20 28 21
0 1 2 3 4 5 6 7 8 9 10 11 INCHES.
CAVALRY SABER, MODEL OF 1913.
ASSEMBLED VIEW AND BLADE.
SPRINGFIELD ARMORY, U. S. A., JUNE 24, 1913.
C.C. Shippee
W.C.
Jno. T. Thompson
CLASS 20
DIVISION 28
DRAWING 19
FILE

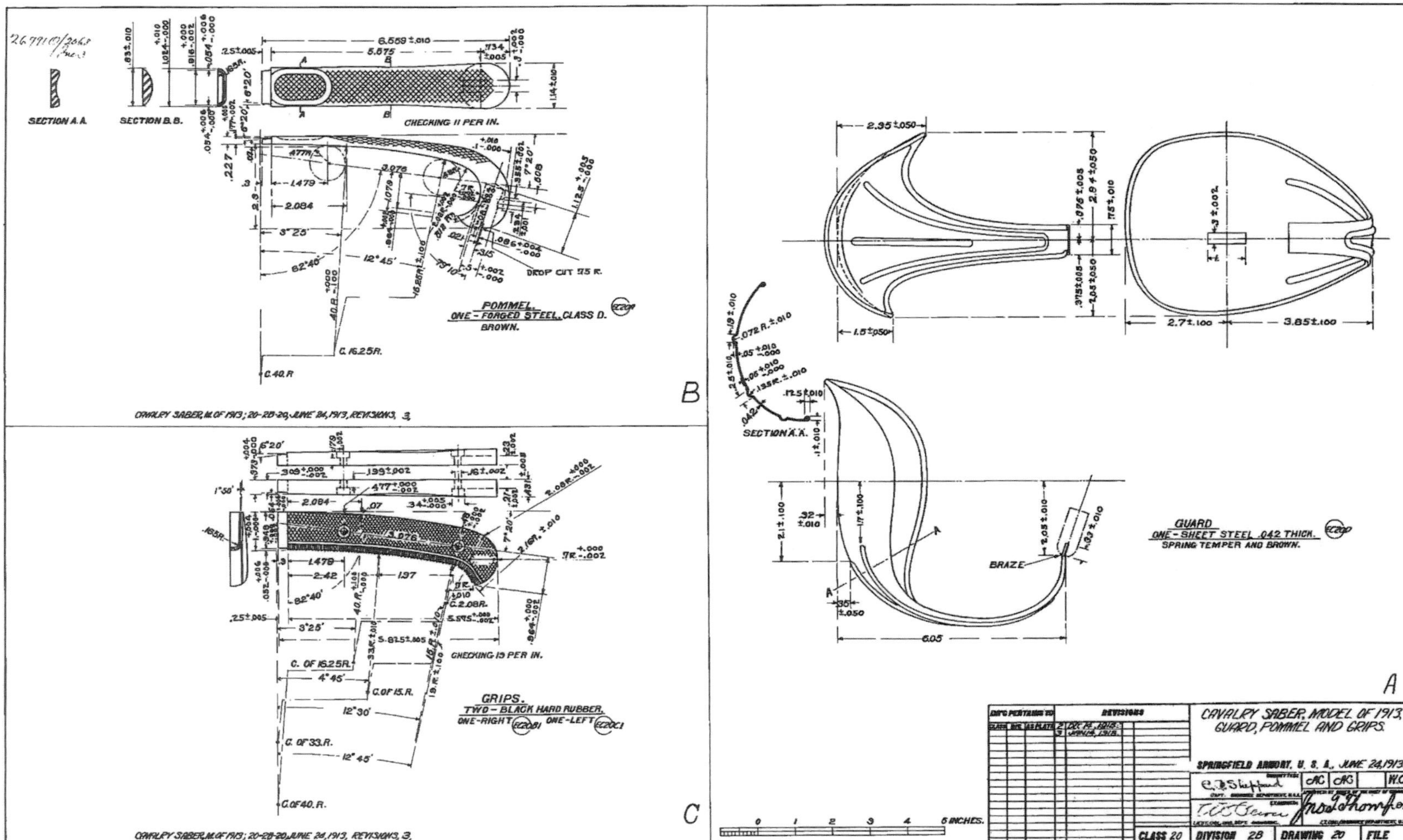
SECTION A.A.
SECTION B.B.
CHECKING 11 PER IN.
DROP CUT 7.5 R.
POMMEL.
ONE - FORGED STEEL, CLASS D.
BROWN.
B
CAVALRY SABER, M. OF 1913; 20-28-20, JUNE 24, 1913, REVISIONS, 3,
CHECKING 15 PER IN.
GRIPS.
TWO - BLACK HARD RUBBER,
ONE-RIGHT ONE-LEFT
C
CAVALRY SABER, M. OF 1913; 20-28-20, JUNE 24, 1913, REVISIONS, 3,
SECTION A.A.
BRAZE
GUARD
ONE - SHEET STEEL .042 THICK.
SPRING TEMPER AND BROWN.
A
0 1 2 3 4 5 INCHES.
REVISIONS
CAVALRY SABER, MODEL OF 1913,
GUARD, POMMEL AND GRIPS.
SPRINGFIELD ARMORY, U. S. A., JUNE 24, 1913.
CLASS 20 DIVISION 28 DRAWING 20 FILE

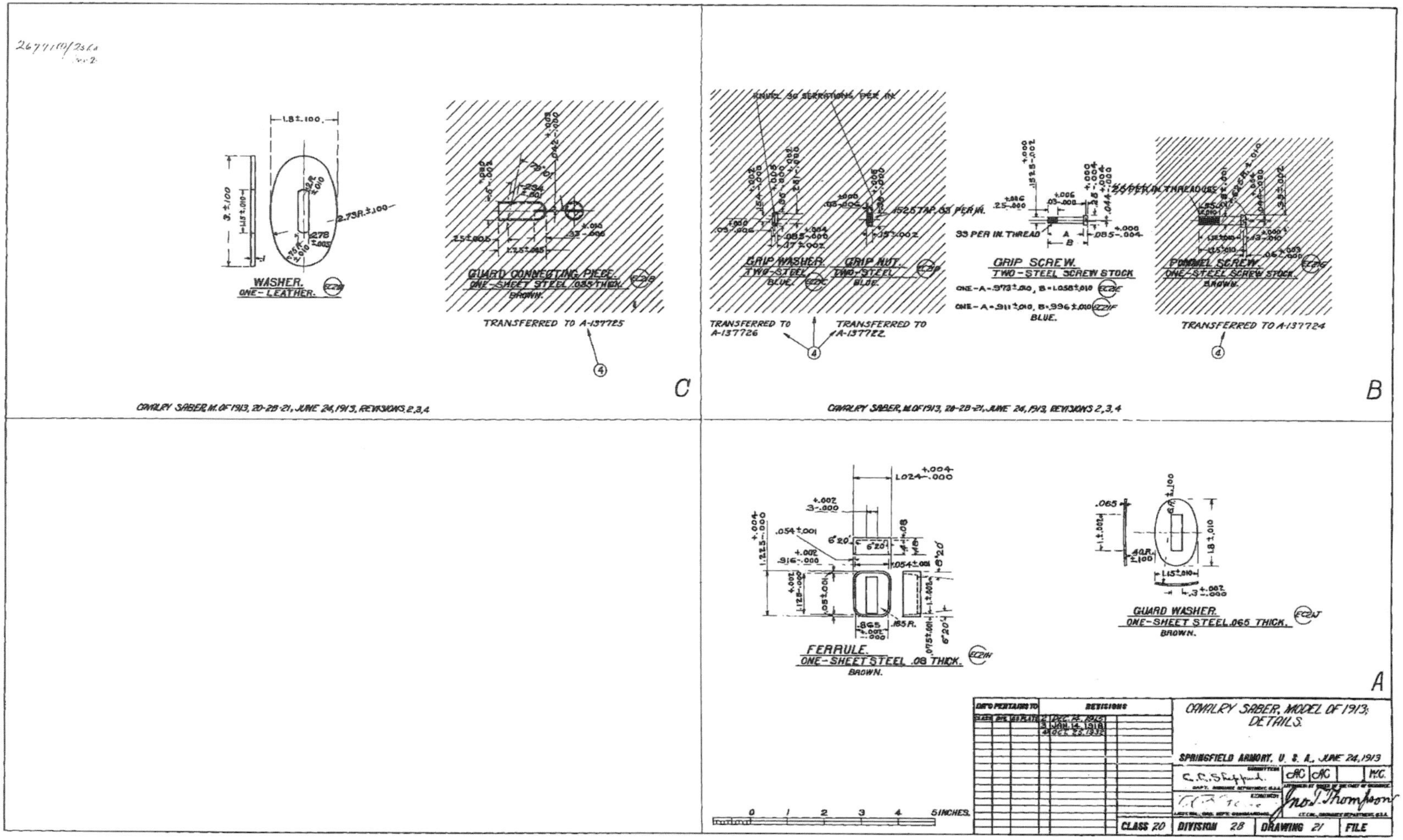

WASHER.
ONE-LEATHER.
GUARD CONNECTING PIECE.
ONE-SHEET STEEL .035 THICK.
BROWN.
TRANSFERRED TO A-137725
C
CAVALRY SABER, M. OF 1913, 20-28-21, JUNE 24, 1913, REVISIONS 2,3,4.
KNURL 30 SERRATIONS PER IN.
GRIP WASHER.
TWO-STEEL.
BLUE.
GRIP NUT.
TWO-STEEL.
BLUE.
33 PER IN. THREAD
GRIP SCREW.
TWO-STEEL SCREW STOCK.
BLUE.
POMMEL SCREW.
ONE-STEEL SCREW STOCK.
BROWN.
TRANSFERRED TO A-137726
TRANSFERRED TO A-137722.
TRANSFERRED TO A-137724
B
CAVALRY SABER, M. OF 1913, 20-28-21, JUNE 24, 1913, REVISIONS 2,3,4
FERRULE.
ONE-SHEET STEEL .08 THICK.
BROWN.
GUARD WASHER.
ONE-SHEET STEEL .065 THICK.
BROWN.
A
5 INCHES.
REVISIONS
CAVALRY SABER, MODEL OF 1913; DETAILS.
SPRINGFIELD ARMORY, U. S. A., JUNE 24, 1913
CLASS 20
DIVISION 28
DRAWING 21
FILE

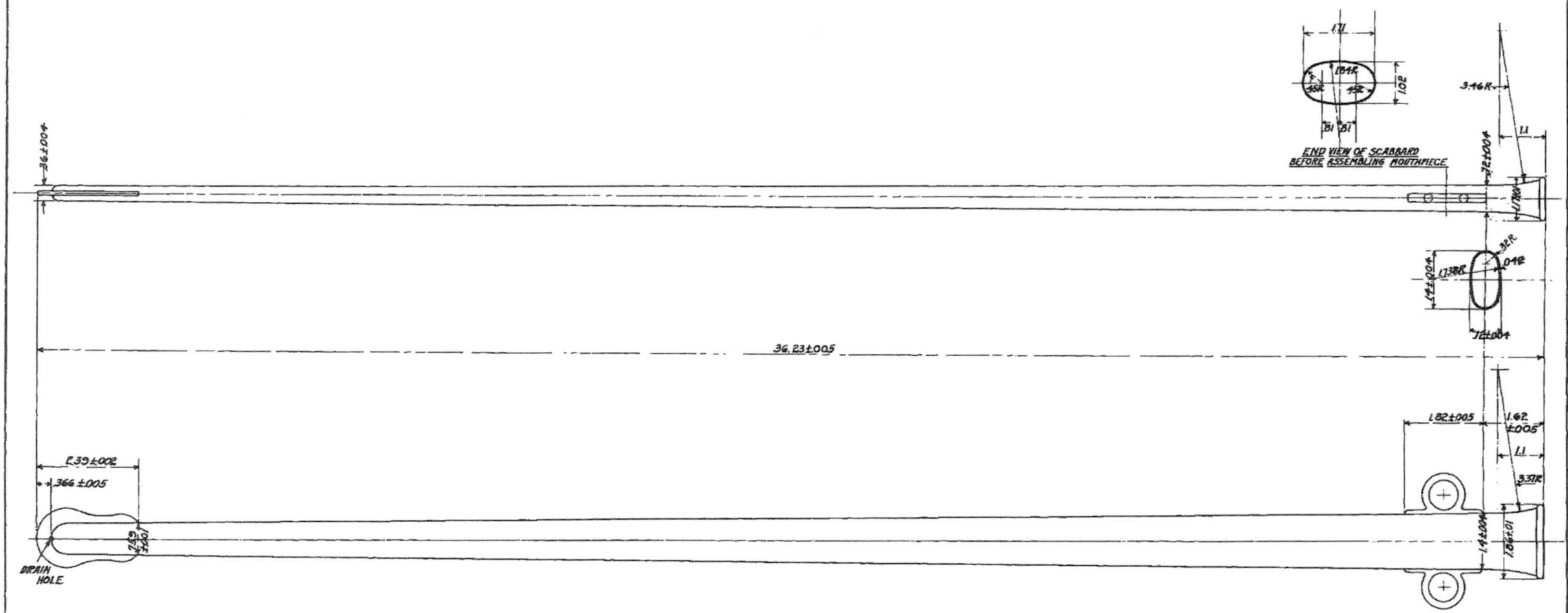

DRAWINGS OF SCABBARD FOR CAVALRY SABER, MODEL OF 1913, FOR OFFICERS IN GARRISON.	CLASS	DIV.	DRG.
ASSEMBLED VIEW	20	28	22
DETAILS	20	28	23

0. 1. 2. 3. 4. 5. 6. 7. 8. 9. 10. 11 INCHES.

SCABBARD FOR
CAVALRY SABER, MODEL OF 1913,
FOR OFFICERS IN GARRISON;
ASSEMBLED VIEW.
SPRINGFIELD ARMORY, U. S. A., NOV. 10, 1913.

C.C. Sheppard, CAPT., ORDNANCE DEPARTMENT, U.S.A. — H.C.M. H.C.M. H.C.M.

Jno. T. Thompson, LT. COL., ORDNANCE DEPARTMENT, U.S.A.

DRG PERTAINS TO — REVISIONS: 2. DEC. 14, 1916

CLASS 20 | DIVISION 28 | DRAWING 22 | FILE

PLATE

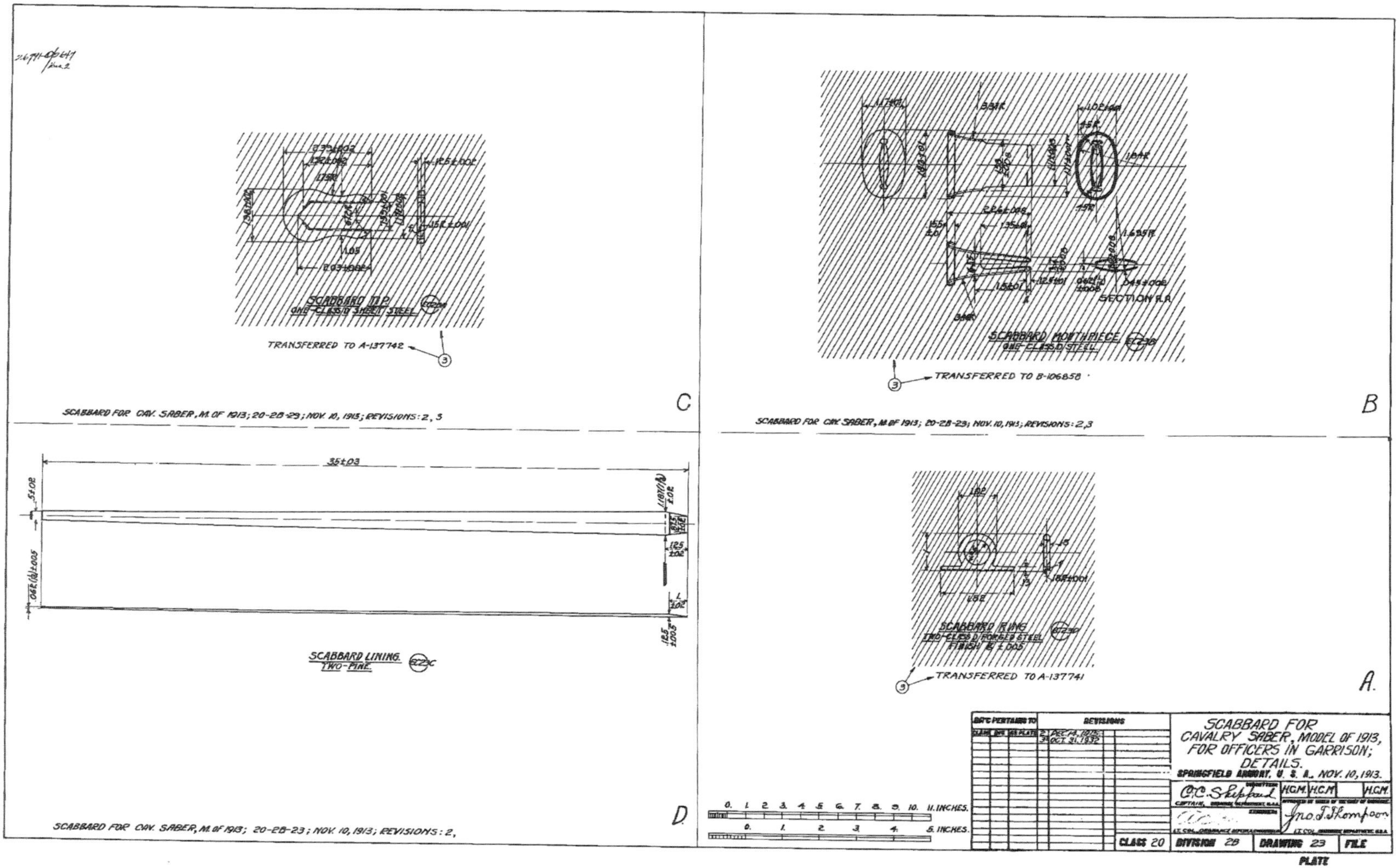
SCABBARD TIP
ONE-CLASS D SHEET STEEL
TRANSFERRED TO A-137742
C
SCABBARD FOR CAV. SABER, M. OF 1913; 20-28-23; NOV. 10, 1913; REVISIONS: 2, 3
SCABBARD MOUTHPIECE
ONE-CLASS D STEEL
SECTION A.A
TRANSFERRED TO B-106858
B
SCABBARD FOR CAV. SABER, M. OF 1913; 20-28-23; NOV. 10, 1913; REVISIONS: 2, 3
35±.03
SCABBARD LINING.
TWO-PINE.
D
SCABBARD FOR CAV. SABER, M. OF 1913; 20-28-23; NOV. 10, 1913; REVISIONS: 2,
SCABBARD RING
TWO-CLASS D FORGED STEEL
TRANSFERRED TO A-137741
A.
REVISIONS
SCABBARD FOR
CAVALRY SABER, MODEL OF 1913,
FOR OFFICERS IN GARRISON;
DETAILS.
SPRINGFIELD ARMORY, U. S. A., NOV. 10, 1913.
CLASS 20
DIVISION 28
DRAWING 23
FILE
PLATE

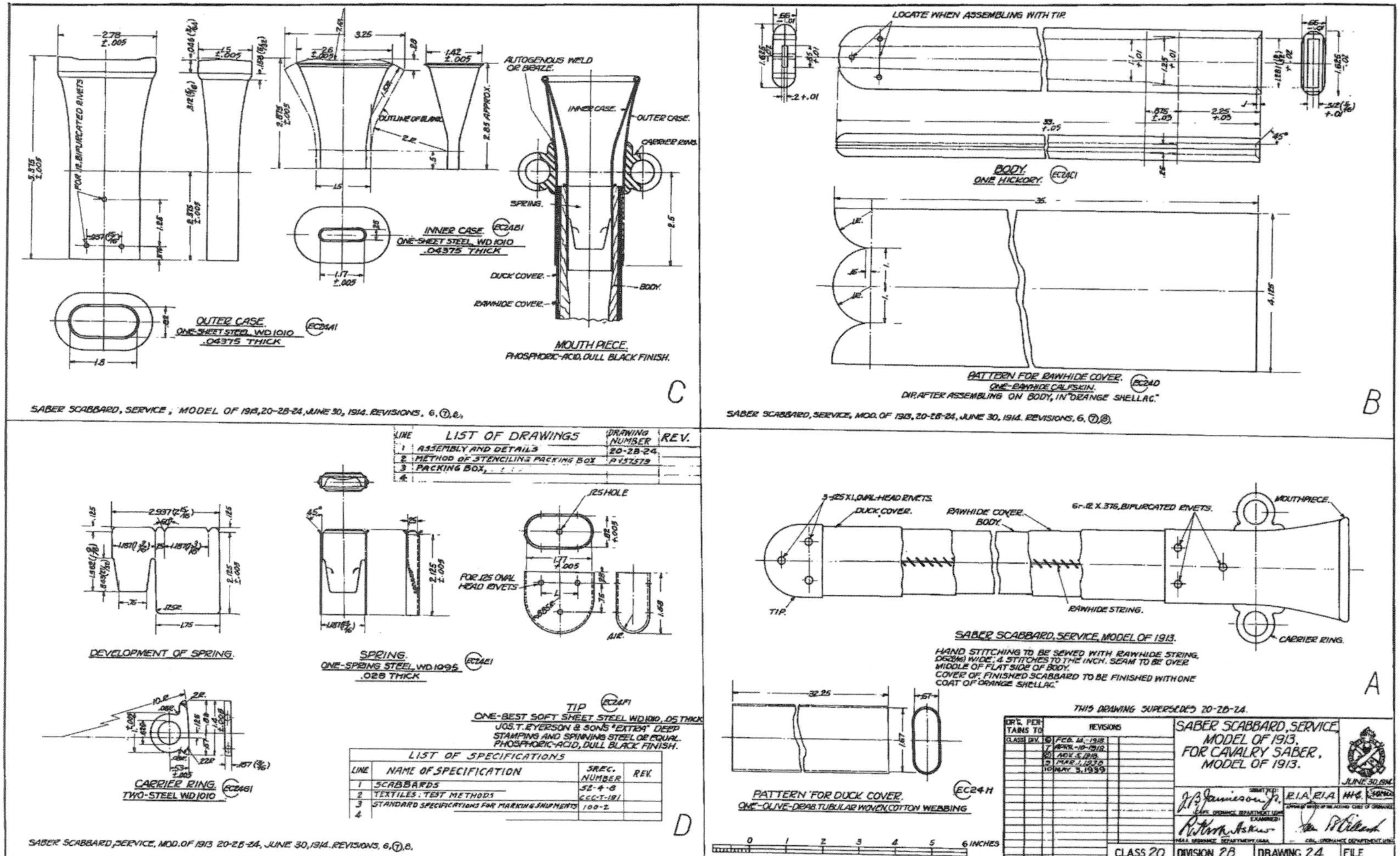

OUTER CASE.
ONE-SHEET STEEL, WD 1010
.04375 THICK
FOR 12 BIFURCATED RIVETS
INNER CASE.
ONE-SHEET STEEL, WD 1010
.04375 THICK
OUTLINE OF BLANK
2.85 APPROX.
AUTOGENOUS WELD OR BRAZE.
INNER CASE.
OUTER CASE.
CARRIER RING.
SPRING.
DUCK COVER.
BODY.
RAWHIDE COVER.
MOUTH PIECE.
PHOSPHORIC-ACID, DULL BLACK FINISH.
C
SABER SCABBARD, SERVICE, MODEL OF 1913, 20-28-24, JUNE 30, 1914. REVISIONS, 6, 7, 8.
LOCATE WHEN ASSEMBLING WITH TIP.
BODY.
ONE HICKORY.
EC24C1
PATTERN FOR RAWHIDE COVER.
ONE-RAWHIDE CALFSKIN.
DIP, AFTER ASSEMBLING ON BODY, IN "ORANGE SHELLAC."
EC24D
B
SABER SCABBARD, SERVICE, MOD. OF 1913, 20-28-24, JUNE 30, 1914. REVISIONS, 6, 7, 8.
LIST OF DRAWINGS
LINE | DRAWING NUMBER | REV.
1 ASSEMBLY AND DETAILS 20-28-24
2 METHOD OF STENCILING PACKING BOX P157579
3 PACKING BOX
DEVELOPMENT OF SPRING.
SPRING.
ONE-SPRING STEEL, WD 1095
.028 THICK
EC24E1
.125 HOLE
FOR .125 OVAL HEAD RIVETS
AIR.
TIP
EC24F1
ONE-BEST SOFT SHEET STEEL WD 1010, .05 THICK
JOS. T. RYERSON & SONS "EXTRA" DEEP STAMPING AND SPINNING STEEL OR EQUAL.
PHOSPHORIC-ACID, DULL BLACK FINISH.
CARRIER RING.
TWO-STEEL WD 1010
EC24G1
LIST OF SPECIFICATIONS
LINE | NAME OF SPECIFICATION | SPEC. NUMBER | REV.
1 SCABBARDS 52-4-8
2 TEXTILES: TEST METHODS CCC-T-191
3 STANDARD SPECIFICATIONS FOR MARKING SHIPMENTS 100-2
D
SABER SCABBARD, SERVICE, MOD. OF 1913 20-28-24, JUNE 30, 1914. REVISIONS, 6, 7, 8.
3-.125 X 1, OVAL-HEAD RIVETS.
DUCK COVER.
RAWHIDE COVER.
BODY.
6-.12 X .375, BIFURCATED RIVETS.
MOUTHPIECE.
TIP.
RAWHIDE STRING.
CARRIER RING.
SABER SCABBARD, SERVICE, MODEL OF 1913.
HAND STITCHING TO BE SEWED WITH RAWHIDE STRING, .0625(IN) WIDE; 4 STITCHES TO THE INCH. SEAM TO BE OVER MIDDLE OF FLAT SIDE OF BODY.
COVER OF FINISHED SCABBARD TO BE FINISHED WITH ONE COAT OF "ORANGE SHELLAC."
A
PATTERN FOR DUCK COVER.
ONE-OLIVE-DRAB TUBULAR WOVEN COTTON WEBBING
EC24H
0 1 2 3 4 5 6 INCHES
THIS DRAWING SUPERSEDES 20-28-24.
REVISIONS
SABER SCABBARD, SERVICE, MODEL OF 1913, FOR CAVALRY SABER, MODEL OF 1913.
JUNE 30, 1914
CLASS 20
DIVISION 28
DRAWING 24
FILE

Appendix 2

Comparison of the Model 1902 Saber for All Officers with the Model 1913 Cavalry Saber

The following is taken from the US army's *Handbook of Ordnance Data*, 1918.

XVII. MISCELLANEOUS EQUIPMENT.

Cavalry saber, model 1913.—This is a straight Cavalry saber with heavy metal guard, as shown in the accompanying illustration. The specifications for the United States Cavalry saber call for the following weight, dimensions, material, and test:

Weight	2.125 pounds
Length of blade	35 inches
Overall length	41½ inches
Material	Simple open-hearth carbon steel

In order to determine the temper and quality of the blade, it is subjected to a bending and striking test, which must be satisfactorily made. This saber is illustrated in detail in engineering drawing, class 20, division 2B, drawing 19.

Officer's saber.—The officer's saber has a curved etched blade, is made in three lengths, with weights as specified below:

Weight:	Pounds.
30-inch saber with scabbard	2.18
30-inch saber without scabbard	1.375
32-inch saber with scabbard	2.25
32-inch saber without scabbard	1.4
34-inch saber with scabbard	2.31
34-inch saber without scabbard	1.44

The blade is made of heat-treated simple carbon steel and the tang is of soft iron. All blades are subjected to a bending and striking test to determine the temper and quality. This saber is illustrated in detail in engineering drawing, class 20, division 2B, drawing 15.

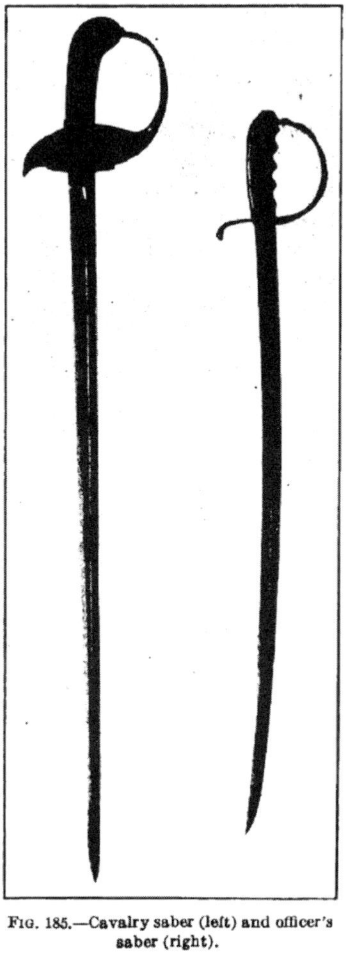

FIG. 185.—Cavalry saber (left) and officer's saber (right).

Appendix 3

Examples of Swords that are not Weapons but Articles of a Dress Uniform

As discussed in the introduction, the Model 1913 Cavalry Saber was the last sword developed as a combat weapon by a major power for its army. However, many nations continued and continue to this day to produce swords as articles of their dress uniforms. The two Luftwaffe dress swords illustrated below are examples of swords designed after 1913 as articles of a military uniform, not intended for use in combat.

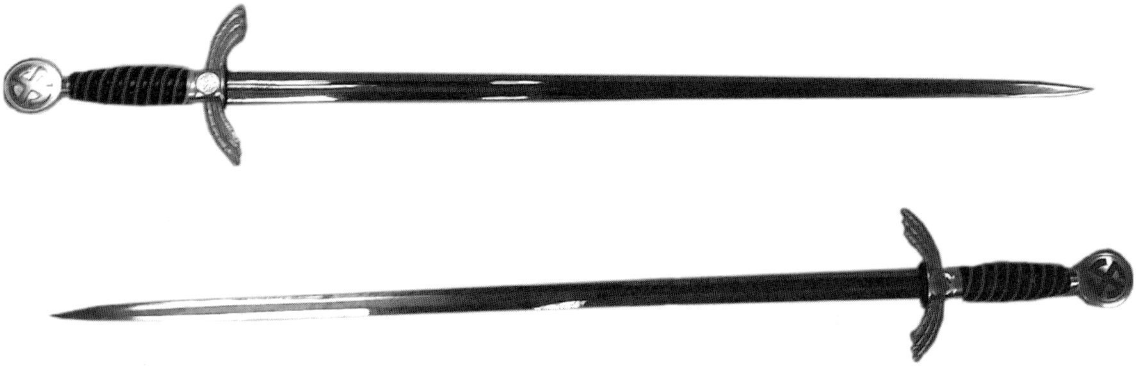

Luftwaffe dress swords of the Nazi regime. (*Courtesy of the Smithsonian Institution, National Museum of American History*)

Glossary

Broadsword An imprecise term referring to any sword with a broad blade, such as a medieval arming sword, but since the eighteenth century, commonly used to refer to basket-hilted swords and the long straight swords carried by heavy cavalry.

Carbine A short rifle, occasionally using the same ammunition as the standard infantry rifle, mostly carried by cavalry and mounted artillery because a shorter weapon was handier to use from the saddle. In the United States, after the Civil War, both the Trapdoor Springfield and the Krag carbines were the primary models. In the final days of the US cavalry, troopers carried the same Model 1903/06 Rifle as the infantry, as its length was not overly cumbersome.

Carabiniers In the Napoleonic Army, heavy cavalry, called carabiniers because of the rifled carbines they were armed with under Louis XIV. Beginning in 1809–10, the carabiniers were armored like the cuirassiers, with cuirasse and steel helmet. They carried a heavy thrusting sword similar to those carried by the cuirassiers.

Chasseur French for a hunter. the Chasseurs à Cheval were French light cavalry, but less elite than the hussars.

Cuirass An armored breastplate and, in most countries, accompanied by an armored backplate. The Austrians at Ratisbon wore only armored breastplates and suffered heavy losses from wounds to the back and sides.

Cuirassiers Heavy armored cavalry. In the sixteenth and early seventeenth centuries, cuirassiers were fully armored from head to knee. By the Napoleonic Wars the armor consisted of a cuirasse and a steel helmet, and they were armed with heavy, long straight double-edged swords.

Cutlass A short curved, broad-bladed sword, usually single-edged and marked by having a guard with good hand protection. Although usually associated with naval weapons, these robust no-nonsense swords were a popular infantry weapon with the colonists of North

America, because of their durability, good hand protection, and general functionality in combat. The sword in Figure 18 above could have been termed a 'cutlass' by colonists of the period.

Dragoons Originally a term for mounted infantry that came to mean heavy cavalry. Napoleon's Dragoons were armed with straight swords, pistols, and muskets. The first cavalry unit in the standing army of the United States was the 2nd Dragoons, who fought with distinction during the Mexican-American War.

Hussars Originally irregular Eastern European light cavalry, specializing in 'small war' operations instead of pitched battles. Louis XIV authorized the first French hussars, Hungarians who had unsuccessfully revolted against Austrian rule. During the French Revolution and even more so under Napoleon, hussar units covered themselves in military glory; between 1792 and 1801, the First Hussars fought in 37 battles, 168 lesser engagements, 1,040 skirmishes, capturing 26,300 enemy soldiers, 40 battle flags and 300 cannons. During the American Civil War, some municipalities raised hussar units, but unlike another French import, the Zouaves, hussars never caught on in America.

Pommel The diminutive of 'pomme', the French word for apple, little apple; the rearmost part of the grip of a sword.

Rapier Of all the swords made to be wielded with one hand, the rapier was the longest. Double-edged with a narrow blade terminating in a sharp point, the rapier was primarily a thrusting weapon, but it could be used to cut. The word 'rapier' is from the Spanish 'espada de robera', or 'the sword of the robe' suggesting its civilian adoption, but there were many that were made for war (see Figure 4 above as an example).

Ricasso The square or rectangular section of steel at the base of the blade abutting the guard.

Notes

Introduction

1. Robert H. Patton, p. 163.
2. Belote, p. 86.
3. Peterson (1991), p. 16.
4. Burton, p. 123.

Chapter 1

1. Bishop, p. 74.
2. Edge and Paddock, p. 26.
3. Ibid., pp. 26–7.
4. Duggan, p. 31.
5. Ibid., p. 138.
6. Nickel, p. 14.
7. Blair, p. 2.

Chapter 2

1. Kennedy, p. 21.
2. Clausewitz, p. 289.
3. Ibid.
4. Ibid., p. 170.
5. Ibid.
6. Ibid., p. 289.
7. Ibid.
8. Krenn and Karcheski, pp. 6–12.
9. Withers, p. 53.
10. Blair, p. 12.
11. Ibid.
12. Burton, pp. 129 and 130.
13. Blair.

Chapter 3

1. Clausewitz, p. 332.
2. Ibid., p. 220.
3. Ibid., pp. 285 and 286.
4. Elting, pp. 698–9.
5. Ibid., p. 232.
6. Ibid., p. 22.
7. Clausewitz, p. 290.

8. Elting, p. 230.
9. Butler (trans.), Vol. 1, p. 379.
10. Elting, p. 229.
11. Ibid., p. 490.
12. Butler (trans.), p. 127.
13. Ibid., Vol. 1, p. 56.
14. Ibid., Vol. 2, p. 219.
15. Ibid., Vol. 2, p. 236.
16. Ibid., Vol. 2, p. 25.
17. Elting, p. 163.
18. Ibid., p. 144.
19. Butler (trans.), Vol. 1, p. 258.
20. Elting, p. 124.
21. George S. Patton, 'A Defense of the Saber', *Journal of the US Cavalry Association*, Vol. 27 (July 1916), p. 48.

Chapter 4

1. https://pooleysword.com/en.
2. Burton, p. 127.
3. Ibid.

Chapter 5

1. Drake, p. 208.
2. Ibid., p. 209.
3. Orr, p. 27.
4. Kredel and Todd, plate 3.
5. Peterson (1991), p. 16.
6. Bass, p. 129.
7. Peterson, 1991, pp. 24–9.
8. Hamilton, p. 19.
9. Ibid., p. 30.
10. Crouch, p. 22.
11. Peterson, 1991, p. 126.
12. Withers, p. 61.
13. Ibid., p. 32.
14. Eisenhower, p. 219.
15. Grant, p. 45.
16. Clary, p. 109.
17. Grant, p. 50.
18. Clary, p. 112.
19. Grant, p. 50.

Chapter 6

1. Crouch, p. 27.
2. Hamilton, p. 17.
3. Ibid., p. 27.
4. Ibid., p. 16.

5. Ibid., p. 160.
6. Sorrel, pp. 23 and 24.
7. Von Borcke, pp. 24 and 25.
8. Sorrel, p. 66.
9. McClellan.
10. Von Borcke, p. 80.
11. Freeman, Vol. 3, p. 31.
12. Von Borcke, p. 414.
13. Freeman, Vol. 3, p. 31.
14. McClellan, p. 277.
15. Freeman, Vol. 3, p. 32.
16. Grant, p. 455.
17. Curry.
18. Belote, pp. 84 and 85.
19. Blair, p. 18.
20. Ibid.
21. Crouch, p. 3.
22. English Patents of Inventions, Specifications: 1863, 283–366, https://books.google.com/books?id=UMstl22Y5BUC&pg=PA19-IA7&lpg=PA19-IA7&dq=british+patent+no.+296+Feb.+2,+1863&source=bl&ots=6vdwDB8w6x&sig=ACfU3U0ezeVrtHwW3qtZZqPk38JwIVToMQ&hl=en&sa=X&ved=2ahUKEwjdoOrQj9DoAhUMj3IEHcVyD90Q6AEwAXoECAIQAQ#v=onepage&q=british%20patent%20no.%20296%20Feb.%202%2C%201863&f=false.
23. https://www.youtube.com/watch?v=XtpMAUf65UQ.

Chapter 7

1. Zimmerman, p. 8.
2. Ibid., p. 92.
3. Silbey, p. 34.
4. Zimmerman, p. 243.
5. Jones, p. 76.
6. Silbey, p. 48.
7. Ibid., p. 64.
8. Jones, p. 216.
9. Ibid., p. 176.
10. Preston, p. 10.
11. Forsyth, p. 7.
12. Ibid., p. 11.
13. Belote, pp. 86 and 87.
14. Annual Reports of the War Department, Vol. IX, Report of the Chief of Ordnance, p. 123.
15. Ibid.
16. Farrington, p. 254.
17. Annual Reports of the War Department, Vol. IX, Report of the Chief of Ordnance, p. 123.
18. Annual Reports of the War Department 1905.
19. US Army Provisional Regulations for Saber Exercise, https://babel.hathitrust.org/cgi/pt?id=hvd.32044080700149&view=1up&seq=7.

20. Hamilton, p. 207.
21. Commissioner of Internal Revenue Annual Report 1914, https://www.irs.gov/pub/irs-soi/1914dbfullar.pdf.
22. Doris.
23. Report of the Chief of Ordnance, War Department Annual Report, 1912, Vol. 1, p. 910.
24. Report of the Cavalry Equipment Board, US Army, Rock Island Arsenal, NARA RG 156.
25. Farrington, pp. 295–7.

Chapter 8

1. Robert H. Patton, pp. 15–17.
2. Ibid., p. 59.
3. Blumenson, *The Patton Papers*, Vol. 1, p. 3.
4. Ibid.
5. Ibid.
6. Robert H. Patton, p. 163.
7. Ibid., p. 93.
8. Nadi, p. 22.
9. Fare.
10. Nadi, p. 20.
11. Report to the Adjutant General on the Fifth Olympic Games, Sept 19, 1912, George S. Patton Papers, Manuscript Division, Library of Congress.
12. Blumenson, *The Patton Papers*, Vol. 1, p. 230.
13. Swedish Olympic Committee, p. 656.
14. Blumenson, *The Patton Papers*, Vol. 1, p. 232.
15. Library of Congress, Manuscript Division, Patton Papers, Report on Olympic Games, p. 4.
16. Blumenson, *The Patton Papers*, Vol. 1, p. 230.
17. Library of Congress, Manuscript Division, Patton Papers, Report on Olympic Games, p. 12.
18. Swedish Olympic Committee, p. 651.

Chapter 9

1. Library of Congress, Manuscript Division, Patton Papers, Report on Olympic Games p. 7.
2. Ibid., pp. 7–8.
3. Ibid., pp. 14–15.
4. Ibid., p. 17.
5. Robert H. Patton, p. 164.
6. Letter to Walter Thorn from GSP on staff of the Army Chief of Staff, Feb. 11, 1913, George S. Patton Papers, Manuscript Division, Library of Congress, Washington DC.
7. Library of Congress, Manuscript Division, Patton Papers, Report on Olympic Games, pp. 16–17.
8. Blumenson, *The Patton Papers*, Vol. 1, p. 245.
9. United States Archives RG 156 e29 26791 0 2482.G56 e.29.
10. Farrington, p. 315.

11. United States Archives RG 153 e29 26791 0 2059.
12. *Springfield Republican*, February 13, 1913.
13. Farrington, pp. 310–11.
14. *Springfield Republican*, February 13, 1913.
15. *Springfield Republican*, February 18, 1913.
16. Blumenson, *The Patton Papers*, Vol. 1, p. 248.
17. *Springfield Union*, March 3, 1913.
18. Blumenson, *The Patton Papers*, Vol. 1, p. 249.
19. War Department Annual Report 1914, https://babel.hathitrust.org/cgi/pt?id=mdp.39015035045973;view=1up;seq=757, p. 467.
20. War Department Annual Report 1914, p. 490.
21. Robert H. Patton, p. 164.
22. Blumenson, *The Patton Papers*, Vol. 1, p. 326.
23. Ibid., pp. 329–39.
24. Ibid., p. 327.
25. Robert H. Patton, p. 164.
26. Patton, 'A Defense of the Saber', *Journal of the US Cavalry Association*, Vol. 27 (July 1916).
27. *Saber Exercise 1914*.
28. Ibid.
29. Blumenson, *The Patton Papers*, Vol. 1, p. 877.
30. Ibid., p. 703.
31. Blumenson, *The Patton Papers*, Vol. 2, p. 21.
32. Ibid., p. 22.

Chapter 10

1. Blumenson, *The Patton Papers*, Vol. 1, p. 213.
2. Ibid., pp. 387–90.
3. Macksey and Batchelor, pp. 14–26.
4. Blumenson, *The Patton Papers*, Vol. 1, p. 446.
5. Ibid., p. 485.
6. Ibid., p. 571.
7. Manchester, p. 101.
8. Ibid., p. 102.
9. Blumenson, *The Patton Papers*, Vol. 1, pp. 609–15.
10. Reed Museum (Second Cavalry) museum guide.
11. *American Armies and Battlefields in Europe*, American Battle Monument Commission, USGPO 1939.
12. Charles Romeyn, 'A Lieutenant Colonel, General Staff Corps "THE SABER, A Reply to General Dickman"', *Journal of the US Cavalry Association*, Vol. 30 (1921).
13. Blumenson, *The Patton Papers*, Vol. 1, p. 483.
14. Dickson and Allen, p. 29.
15. Robert H. Patton, p. 211.
16. Blumenson, *The Patton Papers*, Vol. 1, p. 896.
17. Ibid.

Chapter 11

1. Crouch, p. 39.
2. Blumenson, *The Patton Papers*, Vol. 1, p. 946.
3. Ibid., p. 929.
4. Ibid., p. 943.
5. Ibid., p. 949.
6. Rohde.
7. Belote, p. 85.
8. Sisson, p. 23.
9. Robert H. Patton, pp. 279–81.
10. Blumenson, *The Patton Papers*, Vol. 2, p. 397.

Chapter 12

1. War Department Annual Reports, 1914, Vol. 1, p. 490, available through the Hathi Trust, https://babel.hathitrust.org/cgi/pt?id=uiug.30112001606075&view=1up&seq=506.
2. Farrington, pp. 323, 363, and 364.
3. *Handbook of Ordnance Data*, US Army Ordnance Department, November 15, 1918, available through the Hathi Trust, https://babel.hathitrust.org/cgi/pt?id=uc1.$b74069&view=1up&seq=5.

Bibliography

American Armies and Battlefields in Europe, American Battle Monument Commission, USGPO 1939

Bass, Robert, D., *Swamp Fox, The life and campaigns of General Francis Marion*, Henry Holt and Company, New York, 1959

Belote, Theodore T., *American and European Swords in the Historical Collections of the United States National Museum*, United States Government Printing Office, Washington DC, 1932

Bishop, M.C., *The Gladius*, Osprey Publishing, Oxford, 2016

Blair, Claude, *European & American Arms*, Bonanza Books, New York, 1962

Blumenson, Martin, *The Patton Papers, Vol. 1: 1885–1940*, Houghton Mifflin, Boston, 1974

Blumenson, Martin, *The Patton Papers, Vol. 2: 1940–1945*, Houghton Mifflin, Boston, 1974

Borcke, Heros von, *The Confederate War for Independence, A Prussian Officer with J.E.B. Stuart in Virginia*, J.S. Sanders & Company, Nashville, 1999

Burton, Richard F., *The Book of the Sword*, Dover Books, 1987 (repr., originally published 1894)

Butler, Arthur John (trans.), *The Memoirs of the Baron De Marbot*, Greenhill Books, London, 1990 (repr. of the 1892 edition, available in the common domain at: https://babel.hathitrust.org/cgi/pt?id=uiug.30112114889295&view=1up&seq=11&skin=2021)

Clary, David A., *Eagles and Empire: The United States, Mexico, and the Struggle for a Continent*, Bantam Books, a division of Random House, New York, 2000

Clausewitz, Carl von, *On War*, trans. Michael Howard and Peter Paret, Princeton University Press, Princeton, New Jersey, 1976

Crouch, Howard R., *Historic American Swords*, SCS Publications, Fairfax, Virginia, 1999

Curry, W.L., 'Kilpatrick's Raid Around Atlanta, August 1864', *Journal of the United States Cavalry Association*, 1909–10

Dickson, Paul and Allen, Thomas B., *The Bonus Army, An American Epic*, Walker and Company, New York, 2004

Doris, Lillian, *The American Way in Taxation*, Prentice-Hall, Englewood Cliffs, New Jersey, 1963

Drake, Samuel Adams, *The Making of New England, 1580–1643*, Charles Scribner & Sons, New York, 1920

Duggan, Alfred, *The Story of the Crusades*, Image Books, a division of Doubleday & Co. Inc. Garden City, New York, 1966

Edge, David and Paddock, John Miles, *Arms and Armor of the Medieval Knight: An Illustrated History of Weaponry in the Middle Ages*, Crescent Books, a division of Crown Publishers, New York, 1988

Eisenhower, John, S.D., *Agent of Destiny, The Life and Times of General Winfield Scott*, The Free Press, a division of Simon & Schuster, New York, 1997
Elting, John R., *Swords Around a Throne, Napoleon's Grande Armée*, The Free Press, a division of Macmillan, Inc., New York, 1988
Fare, Malcolm, 'The Development of Fencing Weapons', National Fencing Museum (UK), www.fencingmuseum.com
Farrington, Dusan P., *Sword & Sabers of the United States Army, 1867–1918, The Regulation Models*, Mowbray Publishing, Woonsocket, Rhode Island, 2015
Forsyth, Captain W.W, 'American Cavalry in China', *The Journal of the US Cavalry Association*, July 1, 1903 to June 30, 1904, Nos 49 to 52 inclusive, p. 3, et seq.
Freeman, Douglas Southall, *R.E. Lee, A Biography*, 4 vols, Charles Scribner's Sons, New York, 1935
Grant, Ulysses S., *Personal Memoirs*, originally published 1885, Barnes & Noble, New York, 2003
Hamilton, John D., *The Ames Sword Company, 1829–1935*, Mowbray Co., Providence, Rhode Island, 1983
Johnson, Melvin, *Rifles and Machine Guns, A Modern Handbook of Infantry and Aircraft Arms*, William Morrow and Co., New York, 1944
Jones, Gregg, *Honor in the Dust*, New American Library, New York, 2012
Kennedy, Paul, *The Rise and Fall of the Great Powers*, Vintage Books, a division of Random House, New York, 1987
Kredel, Fritz and Todd, Frederick, P., *Soldiers of the American Army, 1775–1954*, Henry Regnery Company, Chicago, 1954
Krenn, Peter and Karcheski, Walter J. Jr, *Imperial Austria: Treasures of Art Arms & Armor from the State of Styria*, Prestel-Verlag, Munich, 1992
Lawton, Harry W and Knox, George, *The Valiant Knights of Daguerre, Selected Critical Essays on Photography and Profiles of Photographic Pioneers*, University of California Press, Berkeley, 1978
Macksey, Kenneth and Batchelor, John H., *Tank, A History of the Armoured Fighting Vehicle*, Ballantine Books, New York, 1971
Manchester, William, *American Caesar, Douglas MacArthur 1880–1964*, Little, Brown and Company, Boston, 1978
McClellan, Henry B., *I Rode With JEB Stuart*, Da Capo, New York, 1994
Nadi, Aldo, *On Fencing*, originally published 1943, Laureate Press, Sunshine, Florida, 1994
Nickel, Helmut, 'Arms and Armor: From the Permanent Collection', *Metropolitan Museum of Art Bulletin*, v. 49, no. 1 (Summer 1991)
Orr, Charles, *The Pequot War; The Contemporary Accounts of Mason, Underhill, Vincent and Gardener*, The Helman-Taylor Company, Cleveland, 1897, courtesy of the Hathi Trust
Patton, George S., *War as I Knew It*, Houghton Mifflin, Boston, 1947
Patton, Robert H., *The Pattons, A Personal History of an American Family*, Crown Books, New York, 1994
Peterson, Harold, *The American Sword 1775–1945*, Ray Biling Arms Books, Philadelphia, 1991
Peterson, Harold, *Arms and Armor in Colonial America*, Bramhall House, New York, 1956
Preston, Diana, *The Boxer Rebellion: The Dramatic Story of China's War on Foreigners that Shook the World in the Summer of 1900*, Walker and Company, New York, 1999

Rodenbough, Theophilus, *From Everglade to Canyon with the Second Dragoons*, https://books.google.com/books?id=PRIaDqsIL4QC&pg=PP4&lpg=PP4&dq=From+Everglade+to+Ca%C3%B1on+with+the+Second+Dragoons,+:+An+Authentic+Account+of+Service+in+Florida,+Mexico,+Virginia,+and+Th+Theophilus+Francis+Rodenbough&source=bl&ots=s6iSwmgJCZ&sig=ACfU3U0BrsxVkAqZqYau9NYMhqnrT7wGLA&hl=en&sa=X&ved=2ahUKEwiFjoDdtK7uAhVm0FkKHY2kCoMQ6AEwCHoECAgQAg#v=onepage&q=From%20Everglade%20to%20Ca%C3%B1on%20with%20the%20Second%20Dragoons%2C%20%3A%20An%20Authentic%20Account%20of%20Service%20in%20Florida%2C%20Mexico%2C%20Virginia%2C%20and%20Th%20Theophilus%20Francis%20Rodenbough&f=false

Rohde, Aleksandra, *In Defense of My Saber, A Collection of Articles by George S. Patton, Jr.*, Dale Street Books, Silver Spring, Maryland, 2017

Saber Exercise 1914, War Department, Office of the Chief of Staff, Washington DC Government Printing Office, 1914

Silbey, David J., *A War of Frontier and Empire, The Philippine-American War 1899–1902*, Hill and Wang, New York, 2007

Sisson, Frank, *I Marched with Patton: A Firsthand Account of the Second World War Alongside one of the US Army's Greatest Generals*, William Morrow, An Imprint of Harper Collins Publisher, New York, 2020

Sorrel, G. Moxley, *Recollections of a Confederate Staff Officer*, McCowat-Mercer Press, Inc., Jackson, Tennessee, 1958

Swedish Olympic Committee, *The Fifth Olympiad, The Official Report of the Olympic Games of Stockholm 1912*, trans. Edward Adams Ray, Wahlstrom & Widstrand, Stockholm, Sweden, 1913, https://babel.hathitrust.org/cgi/pt?id=mdp.39015015399283&view=1up&seq=9&skin=2021/

Withers, Harvey J.S., *Cavalry Swords of the World 1600–1945*, Harvey Withers Military Publishing, n.p., 2015

Zimmerman, Warren, *First Great Triumph, How Five Americans Made Their Country a World Power*, Farrar, Straus and Giroux, New York, 2003